PELICAN BOOKS

MATHEMATICAL CIRCUS

For more than two decades Martin Gardner's recreational mathematics columns in *Scientific American* have delighted a world-wide audience that includes mathematicians and dreamers, scientists and schoolchildren, computer programmers and poets. He is an occasional critic for *The New York Times Book Review* and the *New York Review of Books*, and a redoubtable amateur magician. He was born in Tulsa, Oklahoma, was educated at the University of Chicago and has lived for many years in a town in New York State on a street named after Euclid.

MARTIN GARDNER

Mathematical Circus

MORE GAMES, PUZZLES, PARADOXES,
AND OTHER MATHEMATICAL ENTERTAINMENTS FROM
SCIENTIFIC AMERICAN

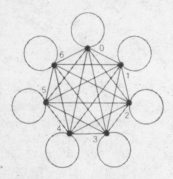

WITH THOUGHTS FROM READERS,
AFTERTHOUGHTS FROM THE AUTHOR AND
105 DRAWINGS AND DIAGRAMS

PENGUIN BOOKS

Penguin Books Ltd, Harmondsworth, Middlesex, England
Penguin Books, 625 Madison Avenue, New York, New York 10022, U.S.A.
Penguin Books Australia Ltd, Ringwood, Victoria, Australia
Penguin Books Canada Ltd, 2801 John Street, Markham, Ontario, Canada L3R 1B4
Penguin Books (N.Z.) Ltd, 182-190 Wairau Road, Auckland 10, New Zealand

First published in the U.S.A. by Alfred A. Knopf, Inc., New York, 1979
First published in Great Britain by Allen Lane 1981
Published in Pelican Books 1981
Reprinted 1982

Most of this book originally appeared in slightly different form
in *Scientific American*

Made and printed in Great Britain by
Richard Clay (The Chaucer Press) Ltd, Bungay, Suffolk

For Donald E. Knuth,
*extraordinary mathematician, computer scientist,
writer, musician, humorist, recreational math buff,
and much more*

Contents

Introduction

*Sometimes these cogitations still amaze
The troubled midnight and the noon's repose.*
—T. S. Eliot*

THE CHAPTERS OF THIS BOOK originally appeared in *Scientific American* as monthly columns with the heading "Mathematical Games." At times mathematicians ask me what I take the phrase to mean. It is not easy to reply. The word "game" was used by Ludwig Wittgenstein to illustrate what he meant by a "family word" that cannot be given a single definition. It has many meanings that are linked together somewhat like the members of a human family, meanings that became linked as the language evolved. One can define "mathematical games" or "recreational mathematics" by saying it is any kind of mathematics with a strong play element, but this is to say little, because "play," "recreation," and "game" are roughly synonymous. In the end one has to fall back on such dodges as defining poetry as what poets write, or jazz as what jazz musicians play. Recreational math is the kind of math that recreational mathematicians enjoy.

Although I cannot define a mathematical game any better than I can a poem, I do maintain that, whatever it is, it is the best way to capture the interest of young people in teaching elementary mathematics. A good mathematical puzzle, paradox, or magic trick can stimulate a child's imagination much faster than a practical application (especially if the application is remote from the child's experience), and if the "game" is chosen carefully it can lead almost effortlessly into significant mathematical ideas.

Not only children but adults too can become obsessed by a puzzle that has no foreseeable practical use, and the history of mathematics is filled with examples of work on such puzzles, by professionals and amateurs alike, that led to unexpected consequences. In his book *Mathematics: Queen and Servant of Science*, Eric Temple Bell cites the early work on classifying and enumerating knots as something that once seemed little more than puzzle play, but later became a flourishing branch of topology:

> So the problems of knots after all were more than mere puzzles. The like is frequent in mathematics, partly because mathematicians have sometimes rather perversely reformulated serious problems as seemingly trivial puzzles abstractly identical with the difficult problems they hoped but failed to solve. This low trick has decoyed timid outsiders who might have been scared off by the real thing, and many deluded amateurs have made substantial contributions to mathematics without suspecting what they were doing. An example is T. P. Kirkman's (1806–1895) puzzle of the fifteen schoolgirls (1850) given in books on mathematical recreations.

Some mathematical puzzles really are trivial and lead nowhere. Yet both types have something in common, and no one has expressed this better than the eminent mathematician Stanislaw Ulam in his autobiography, *Adventures of a Mathematician:*

> With all its grandiose vistas, appreciation of beauty, and vision of new realities, mathematics has an addictive property which is less obvious or healthy. It is perhaps akin to the action of some chemical drugs. The smallest puzzle, immediately recognizable as trivial or repetitive, can exert such an addictive influence. One can get drawn in by starting to solve such puzzles. I remember when the *Mathematical Monthly* occasionally published problems sent in by a French geometer concerning banal arrangements of circles, lines and triangles on the plane. "Belanglos," as the Germans say, but nevertheless these figures could draw you in once you started to think about how to solve them, even when realizing all the time that a solution could hardly lead to more exciting or more general topics. This

is much in contrast to what I said about the history of Fermat's theorem, which led to the creation of vast new algebraical concepts. The difference lies perhaps in that little problems can be solved with a moderate effort whereas Fermat's is still unsolved and a continuing challenge. Nevertheless both types of mathematical curiosities have a strongly addictive quality for the would-be mathematician which exists on all levels from trivia to the more inspiring aspects of mathematics.

MARTIN GARDNER
March 1979

Mathematical
Circus

CHAPTER 1

Optical Illusions

OPTICAL ILLUSIONS—pictures, objects, or events that are not what they seem to be when perceived—have played and still play important roles in art, mathematics, psychology, and even philosophy. The ancient Greeks distorted the columns of the Parthenon so they would look straight to people on the ground. Renaissance muralists often distorted large wall paintings so they would appear normal when viewed from below. Mathematicians are interested in optical illusions because many of them are related to perspective (a branch of projective geometry) and other aspects of geometry. Psychologists investigate illusions to learn how the brain interprets the data that come to it by way of the senses. And philosophers of various schools of direct realism, who maintain that we perceive actual objects external to our minds, have the problem of explaining how errors of perception can arise.

On less serious levels, optical illusions are just plain fun. One enjoys being fooled by them for reasons not unlike those that underlie the delight of being fooled by a magician. Illusions remind us that the big outside world is not always what it seems.

In this chapter we will concentrate on a few optical illusions that are not so well known, and that have strong mathematical flavors.

The process by which the brain interprets visual data is so complex and little understood that it is no surprise to find psychologists disagreeing sharply over explanations for even the simplest illusions. One of the oldest is the apparent increase in the size of the sun, moon, and constellations when they are near the horizon. The late Edwin G. Boring of Harvard University wrote many papers arguing that the "moon illusion" is caused primarily by the raising of one's eyes. A different view, going back to Ptolemy, is defended by Lloyd Kaufman and Irvin Rock in their article on "The Moon Illusion" in *Scientific American*, July 1962. Their "apparent distance" theory is in turn challenged by Frank Restle in a paper in *Science*, February 20, 1970.

Today's approach is to regard most visual illusions as occurring in the brain as it searches its memory for what Richard L. Gregory calls the "best bet": the interpretation that best explains the visual data in terms of the brain's stored experience. Such a view is supported by recent discoveries that many animals, including birds and fish, have illusions that can be explained in this way and also by work with people in cultures that differ markedly from ours. Zulus, for example, live in an almost completely round world. Their huts and doors are rounded. They plow fields in curves. Straight lines and right angles are seldom seen and there is no word for "square" in their language. As John Updike puts it in the second stanza of his poem "Zulus Live in Land Without a Square":

> *When Zulus cannot smile, they frown,*
> *To keep an arc before the eye.*
> *Describing distances to town,*
> *They say, "As flies the butterfly."*

Several recent studies have shown that optical illusions involving parallel lines and angular corners, so common to the rectangular world of technologically advanced societies, are difficult for Zulus to perceive. The philosophers John Locke and George Berkeley both considered the question of whether a man born blind, who suddenly gains his sight, would be able to decide, without touching them, which of two objects was a cube and which a sphere. Locke and Berkeley thought not. Gregory's *Eye and Brain* summarizes recent studies along such lines and, although they are inconclusive, they seem to support both philosophers, again providing evidence for the modern view that most optical illusions are caused by the brain's faulty interpretation of input data.

An amusing new development in visual illusions is the discovery of "undecidable figures": drawings of objects that cannot exist. The brain, unable to make sense of them, is thrown into a strange state of befuddlement. (They are analogous to such undecidable sentences as "This statement is false" or "Don't miss it if you can.") The best-known undecidable figure is the notorious three-pronged (or is it two-pronged?) "blivet," which began circulating among engineers and others in 1964. The March 1965 cover of *Mad* showed a grinning Alfred E. Neuman (with four eyes) balancing the blivet on his index finger. Roger Hayward contributed an article on "Blivets: Research and Development" to *The Worm Runner's Digest* (December 1968) in which he presented several variations [*see Figure 1*].

Another well-known undecidable figure is the square staircase around which one can climb or descend forever without getting higher or lower. It can be seen in Maurits C. Escher's 1960 lithograph "Ascending and Descending" [see my *Mathematical Carnival*, page 95] and in the same artist's 1961 lithograph of a waterfall operating a perpetual motion machine. This mystifying illusion, designed by the British geneticist L. S. Penrose and his son, the mathematical physicist Roger Penrose, was first published in their article "Impossible Objects: A Special Type of Visual Illusion" in *The British Journal of Psychology* (February 1958, pages 31–33).

FIGURE 1

Roger Hayward's "undecidable" monument

The same two authors made use of it in their collection of original "Christmas Puzzles" for *The New Scientist* (December 25, 1958, pages 1580–81). Assuming [*see Figure 2*] that it takes three steps to go from the ground *A* to the top of step *B*, how can one get from *A* to *C* by climbing no more than 10 steps? The solution is possible only because the structure itself is not.

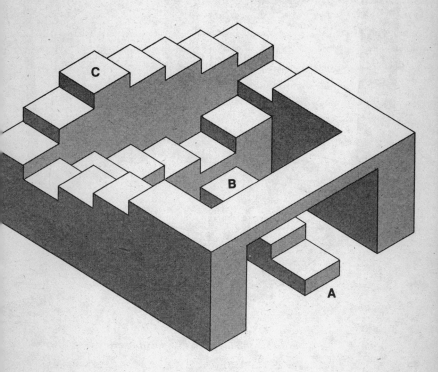

FIGURE 2
A puzzle based on the Penrose stairway

A third familiar impossible object is the skeleton of the cube held by a seated figure in another Escher lithograph that can be

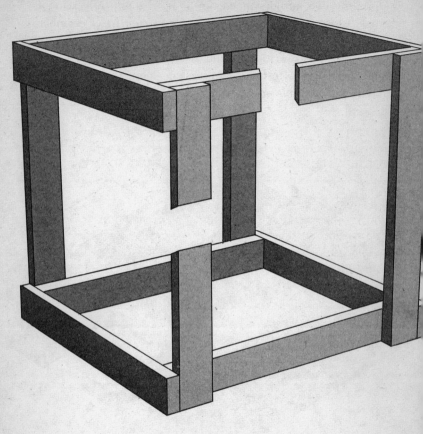

FIGURE 3
Possible model for an impossible crate

seen on page 94 of my *Mathematical Carnival*. The "Letters" department of *Scientific American* reproduced a photograph of such a "Freemish Crate" (as it was called) in its June 1966 issue, but the picture was obtained by doctoring the print. There is, however, a way to build an actual model that will produce a genuine photograph of a Freemish crate. It is explained by William G. Hyzer in *Photo Methods for Industry*, January 1970. Hyzer's model is shown in Figure 3. If the model is

turned and tilted so that one open eye sees the gaps exactly co-
inciding with two back edges of the crate, the brain assumes
that the back edges are in front, producing a mental image of
the impossible cube.

The fact that we have two eyes makes possible many curious
illusions. Hold your extended index fingers horizontally before
your eyes, tips touching. Focus through the fingers on a distant wall and then separate the fingertips slightly. You will see
a "floating hot dog" between the two fingers. It is formed, of
course, by overlapping fingertips, each seen by a different eye.
Another ancient illusion of binocular vision is produced by
holding a tube (such as a rolled-up sheet of paper) to your
right eye like a telescope. Your left hand, palm toward you, is
placed with its right edge against the tube. If you slide the hand
back and forth along the tube, keeping both eyes open and looking at a distant object, you will find a spot where you seem to
be looking through a hole in the center of your left palm.

Under certain circumstances single-eyed vision gives an illusion of depth. Looking at a photograph through a tube with one
eye produces a slight three-dimensional effect. One of the most
striking illusions of monocular vision is shown in Figure 4. The
page must be tilted back until it is almost flat. If the picture is

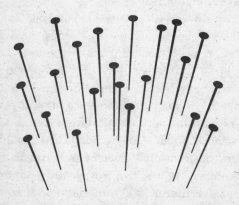

FIGURE 4
The nails that stand up

viewed with one open eye from near the lower edge of the page, close to a spot above the point where all the lines would meet if they were extended downward, in a moment or two the nails will seem to stand upright. William James, in Volume 2, Chapter 19, of his famous *Principles of Psychology*, after giving an excellent explanation of this illusion, adds the following succinct summary of the modern approach to perception: "In other words, we see, as always, the most probable object."

An amazing binocular illusion is called the Pulfrich pendulum after its discoverer, Carl Pulfrich, who first described it in a German periodical in 1922. The pendulum is simply a piece of string, from one foot to four feet long, with a small object tied to one end. Have someone hold the other end and swing the bob back and forth along a plane perpendicular to your line of vision. Stand across the room and view the swinging bob by holding one lens of a pair of sunglasses over one eye. Both eyes must remain open. Keep your attention at the center of the swing rather than following the moving bob. The bob will appear to swing in an elliptical orbit! Shift the dark glass to your other eye and the bob will travel the same elliptical path but in the opposite direction. The depth illusion is so strong that if a large object is held behind the path of the bob, the bob actually seems to pass through it like a ghost.

The Pulfrich illusion is explained by Gregory as arising from the fact that your dark-adapted eye sends messages to your brain at a lower speed than your uncovered eye. This time lag forces your brain to interpret the bob's movement as being alternately in front and in back of the plane in which it swings.

Similar depth illusions are experienced if you look at a television picture with the dark glass over one eye or with one eye looking through a pinhole in a card. When something in the picture moves horizontally, it seems to travel in front of the screen or behind it. It was this illusion that prompted several companies to advertise in 1966 a special pair of spectacles that, according to the advertisements, enabled one to see flat television pictures in three dimensions. The price was high and of course the spectacles were merely inexpensive sunglasses with a piece of transparent plastic for one eye and darkened plastic for the other.

A familiar category of illusions, much analyzed by the Gestalt school of psychology, concerns images capable of two interpretations of equal or nearly equal probability. The mind fluctuates, unable to settle on the best bet. The pattern of cubes that suddenly reverses, so that the number of cubes changes, is perhaps the best-known example. In recent years we all have been annoyed by looking at photographs of lunar craters and finding it difficult not to see them as mesas, particularly if the picture is turned so that the craters are illuminated by sunlight, from below, an angle of lighting that is seldom experienced.

The black vase with contours that can be seen as profiles of two faces, another much-reproduced illusion of fluctuating gestalts, unexpectedly popped up in the new Canadian flag when it was officially adopted in 1965 after months of wrangling in the House of Commons. Direct your attention to the white background at the top of the maple leaf [*see Figure 5*]. You will see

FIGURE 5
The Canadian flag and its two angry men

the profiles of two men (Liberal and Conservative?), foreheads together, snarling (one in English, the other in French?) at each other. Once you have spotted these faces you should have little difficulty understanding the odd-shaped polygons in Figure 6.

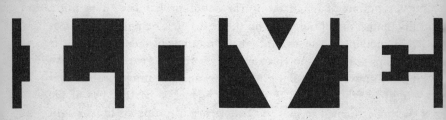

FIGURE 6

A Gestalt puzzle. What do the black shapes represent?

The Necker cube (after L. A. Necker of Switzerland, who wrote about it in the 1830's) is another much-studied figure that reverses while you are looking at it. The Penroses, in their Christmas puzzles mentioned earlier, had the clever idea of adding a beetle to the "cube," in this instance a rectangular box [*see Figure 7*]. The beetle appears to be on the outside. Stare

FIGURE 7

Put the bug inside the box.

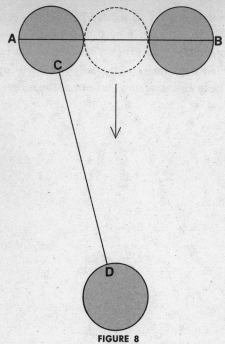

FIGURE 8
An equal-distance illusion

at the back corner of the box and imagine it to be the corner nearest you. The box will suddenly flip-flop, transporting the beetle to the floor *inside*.

A surprising illusion, perhaps related to the Müller-Lyer illusion (two lines of equal length that appear different because of arrow lines that point inward at the ends of one line and outward at the ends of the other) can be demonstrated with three pennies. Place them in a row [*see Figure 8*]. Ask someone to slide the middle penny down until distance *AB* equals distance *CD*. Almost no one slides the coin far enough; indeed, it is hard to believe, until you measure the lines, that this illustration gives the correct position. The trick can also be done with larger coins, circular coasters, water glasses and other similar objects.

The "ghost-penny" illusion, better known to magicians than to psychologists, is illustrated in Figure 9. Hold two pennies between the tips of your forefingers and rub them rapidly back and forth against each other. A ghost penny will appear—but why should it be only at one end and not at the other?

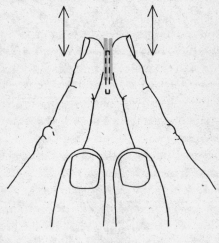

FIGURE 9
The "ghost penny"

ANSWERS

To CLIMB in 10 steps to the top of the Penrose staircase, walk up four steps, turn right, continue up three more steps, go back around the level U-shaped path, down three steps, then up three to the top.

I have never seen in print an explanation of the ghost-penny illusion, but so many readers sent such a convincing explanation that I have no doubt it is correct. When the two pennies are rubbed back and forth in the manner described, the angle of the fingers causes the coins to diverge in their forward positions and create completely separate images. Contrariwise, the slight lateral motion on the V side of the fingers causes their backward positions to converge and overlap. The result is that the separated forward images are faint, but the overlapping backward images reinforce one another to create a single, stronger image.

Readers described many simple ways of confirming this theory. Marjorie Lundquist and S. H. Norris, for example, proposed the following experiment. Turn your palms outward, pointing the thumbs toward you. If two pennies are held between the tips of the thumbs and rubbed together, the ghost coin appears on the V side of the thumbs, away from your body. This is what one would expect from the slight lateral motions that now overlap images on the far side. If the thumbs are held so that instead of making a V they form a straight line, the lateral motions are equalized on both sides and you see two ghost pennies. The same symmetrical double ghost results when the coins are held between the index fingertips, but instead of rubbing forward and back, they are rubbed vertically up and down.

Another striking confirmation of the theory, which I discovered myself, is obtained by rubbing the tips of your forefingers rapidly back and forth with no coins at all between them. The divergence in front and overlap in back is obvious. You will see a ghost fingertip within the V, with the edge of a fingernail right down its center!

The ghost-penny illusion can be presented as a magic trick. Start with a penny (or nickel or quarter) palmed in the right hand. Borrow two similar coins from someone and hold them between the tips of your right thumb and forefinger. Slide the coins rapidly back and forth to create the ghost, holding the hand so the palmed coin can't be seen. After the ghost appears, close your hand into a fist, then open it to show that the ghost has materialized as a third coin.

CHAPTER 2

Matches

PAPER OR WOODEN MATCHES have two properties that lend themselves to mathematical amusements: they can be used as "counters," and they are handy models of unit line segments. A full compilation of mathematical recreations using matches would fill a large volume. In this chapter we consider a few representative samples of tricks, games, and puzzles with matches.

An old trick known to magicians as the "piano trick" (because of the spectator's hand positions) can be presented as a miraculous exchange of odd and even parity. Ask someone to place his hands on the table, palms down. Insert two paper matches between each pair of adjacent fingers except for the ring finger and little finger of one hand, which get only a single match [*see Figure 10*]. Remove the pairs of matches one at a time. Separate the matches of each pair and place them on the table, one match in front of each of the spectator's hands. Each time you do this say, "Two matches." Continue in this way, forming a small pile of matches in front of each hand, until only the single match remains. Take this match from his hand, hold it in the air, and say: "We have here two piles of matches,

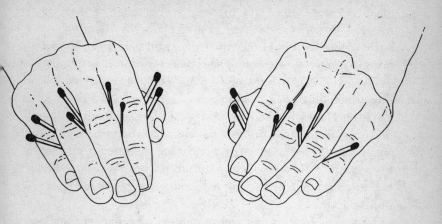

FIGURE 10
The "piano trick"

each formed with pairs. To which pile shall I add the single odd match?" Place the match on the pile designated.

Point to the pile on which you dropped the match and say, "This is now a pile containing an extra match." Point to the other pile and say, "This remains a pile made up of pairs." Wave your hands over both piles and announce that you have caused the odd match to travel invisibly over to the other pile. To prove that this has indeed occurred, "count" the matches in the pile on which you dropped the single match by taking the matches by pairs and sliding them to one side. "Count" is in quotes because you do not actually count them. Instead, merely repeat, "Two matches" each time you move a pair to one side. The pile will consist entirely of pairs, with no extra match left over. "Count" the other pile in the same way. After the last pair has been slid aside a single match will remain. With convincing patter the trick will puzzle most people. Actually it is self-working, and the reader who tries it should easily figure out why.

A trick that goes back to medieval times and can be found in the first compilation ever made of recreational mathematical material, *Problèmes plaisans et délectables*, by Claude Gaspar Bachet, published in France in 1612, is still performed by magicians in numerous variants. The classical version is as follows.

Twenty-four matches are placed on a table, together with any three small objects—say a dime, a finger ring, and a house key. Three spectators are chosen to assist. Designate them 1, 2, and 3. To make sure that this order is remembered (you say), give one match to spectator 1, two matches to spectator 2, and three to spectator 3. These matches are taken from the 24 on the table, leaving a pile of 18. Ask each spectator to put his matches in his pocket.

Turn your back so that you cannot see what happens and ask that spectator 1 take any one of the three objects and put it in his pocket. Spectator 2 picks either of the two remaining objects. The third spectator pockets the last object. Now ask the person who took the dime to remove from the table as many matches as you originally gave him and hold them in a closed fist. (You have no way of knowing who it is because your back is still turned.) Ask the person who took the ring to remove twice as many matches as you originally gave him and hold them in his fist. Ask the one who took the key to take four times the number of matches he was given and hold them.

You turn around and, after a few moments of feigned extrasensory concentration, tell each person which object he took. The clue is provided by the number of matches remaining on the table. There are six possible permutations of the three objects in the pockets of the three spectators. Each permutation leaves a different number of matches on the table. If we designate the objects S, M, and L for small, medium, and large, the chart in Figure 11 shows the permutation that corresponds to each possible remaining number of matches. (Note that it is impossible for four matches to remain. If you see four matches on the table, someone goofed and the trick has to be repeated.)

Matches Left	Spectators		
	1	2	3
1	S	M	L
2	M	S	L
3	S	L	M
5	M	L	S
6	L	S	M
7	L	M	S

FIGURE 11
Key for the three-object trick

Dozens of mnemonic sentences have been devised so that a performer can determine quickly how the three objects are distributed. Bachet labeled the objects *a, e, i,* the first three vowels, and used this French sentence: (1) *Par fer* (2) *César* (3) *jadis* (5) *devint* (6) *si grand* (7) *prince.* The two vowels in each word or phrase provide the needed information. For example, if the magician sees five matches on the table, the fifth word, "*devint,*" tells him that object *e* was taken by the first spectator (who had been given one match) and object *i* by the second spectator (who had two matches); the remaining object, *a,* must be in the pocket of the remaining spectator, who had been given three matches at the start of the trick. Other 17th-century

European tricksters, also using the first three vowels for the objects, remembered the six permutations by the first two vowels of each word in the Latin line: *Salve certa animae semita vita quies*.

For the version given here, with objects designated *S, M, L*, a good English mnemonic sentence was invented by Oscar Weigle, an amateur magician: (1) *Sam* (2) *mo*ves (3) *sl*owly [(4) since] (5) *mu*le (6) *lo*st (7) *lim*b. The first two appearances of the key letters, shown in italics, give the objects taken by the first and second spectators respectively, leaving the third object to be paired with the third spectator. Many other mnemonic sentences for the trick have been published in English and other languages. The reader may enjoy making up one of his own. The objects can be designated by other letters, such as *A, B, C* or *L, M, H* (for light, medium, heavy), or by the initial letters of whatever objects are used, and so on. It is convenient to introduce a dummy fourth word, as in Weigle's sentence, where it is shown bracketed, even though four remaining matches are not possible. This enables the performer to count the matches quickly by repeating the words of the sentence without having to worry about skipping number 4 if there are more than three matches. An interesting extension of the trick, dating from 1893, to *n* players and *n* objects using an *n*-based number system, is given in W. W. Rouse Ball's *Mathematical Recreations and Essays* (page 30 in the revised 1960 edition).

Some elementary number theory and the fact that a new folder of paper matches contains 20 matches lie behind a more recent mind-reading stunt. While your back is turned ask someone to tear from a full folder any number of matches from one through 10 and pocket them. Then have him count the remaining matches, add the two digits in this number and tear from the folder the number of matches equal to the sum. (For instance, if 16 matches remain, he adds 1 and 6 and tears out seven more matches.) These matches he also puts in his pocket. Finally, he tears out a few more matches—as many as he wishes—and holds them in his closed fist. You turn around and take the folder from him, mentally counting the remaining matches as you put the folder in your pocket. You can now tell him the number of matches in his fist. The first two operations always leave nine matches in the folder. (Can you prove that this must be the case?) Therefore you have only to subtract from 9 the number of matches still in the folder to learn the number concealed in his hand.

A variety of "take away" games, such as nim, can be played with matches, and there are various betting games in which matches are used as counters and concealed in a fist. The following game is one in which paper matches are particularly appropriate because of their shape and the fact that they can be obtained with differently colored heads. The game was invented recently by Jurg Nievergelt, a computer mathematician, who calls the game "Hit-and-Run." It is ordinarily played on an order-4 square matrix [*see Figure 12*].

One player starts with a full folder of black-tipped matches, the other with a full folder of gray-tipped matches. It is a pleasant coincidence that 40 matches are just enough. Players take turns placing a single match on any line segment of the matrix. Black's object is to construct a path connecting the two black sides of the board, Gray's is to construct a path connecting the other two sides. (Opposing paths may cross each other at a right angle.) The first player to build his path wins. The game is called Hit-and-Run because a move can block an opponent's path (a hit) and at the same time extend the player's own path (a run).

The game bears a superficial resemblance to Piet Hein's game of Hex and later variants such as Bridg-it and Twixt, but the mathematical structure behind it is quite different. As in tick-tacktoe there is a simple proof that if both players of Hit-and-Run play rationally, the game is either a first-player win or a draw. Assume that the second player has a winning strategy. The first player can steal the strategy by first making an irrel-

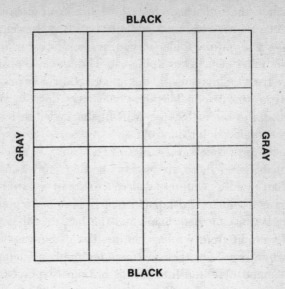

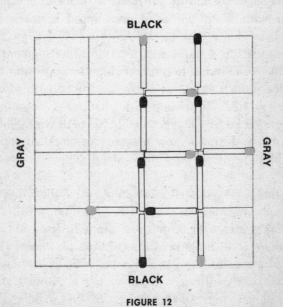

FIGURE 12

Hit-and-Run board (top) and a completed game (bottom)

evant move, and thereafter playing the winning strategy. The irrelevant move can only be an asset, never a liability. If the winning strategy later calls for the irrelevant move, the move has already been made so another irrelevant move is made. In this way the first player can win. Because this contradicts the initial assumption, it follows that the second player has no winning strategy. The first player, consequently, can either win or draw, although the proof provides no information about the strategy he must follow.

On an order-2 square Hit-and-Run is easily seen to be a win for the first player [*at left in Figure 13*]. Black's first move (B1) forces Gray to reply G1. B2 puts Black in position to complete a path by one move in either of two ways (marked B3), and so there is no way Gray can stop Black from winning on the next move. The first player can win in a similar way by playing first on any of the six vertical line segments.

Readers can try to prove that on the order-3 square the game can also always be won by the first player (Black) if he or she

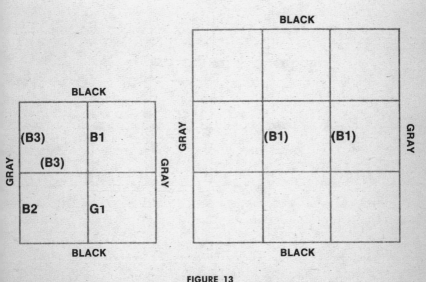

FIGURE 13

First-player win on order-2 (left) and order-3 (right) boards

plays first on either line marked B1 [*at right in Figure 13*]. Nievergelt obtained such a proof by exhausting all possibilities; because it is long and tedious it will not be given. To my knowledge it is not yet known whether Hit-and-Run on the order-4 board, or any square board of a higher order, is a win for the first player or a draw, if both sides play their best.

Black and gray matches can also be used for playing Connecto, a game described by David L. Silverman in his book *Your Move* (McGraw-Hill, 1971). Here too the players alternate in placing matches on a square matrix of any size, but the object now is to be the first to enclose a region of any shape within a boundary of one's own matches. In Figure 14 Black has won the game. Can you discover Silverman's simple strategy by which the second player can always prevent the first player from winning, even on an infinite matrix?

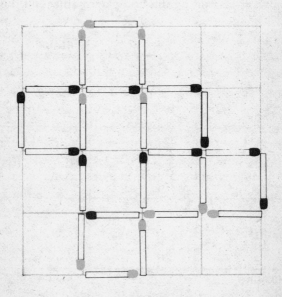

FIGURE 14
Completed game of Connecto

Finally, here are seven entertaining match puzzles [*see Figures 15A and 15B*]:

1. Remove six matches, leaving 10.
2. The six matches are shown forming a map requiring three colors, assuming that no two regions sharing part of a border can have the same color. Rearrange the six to form a planar map requiring four colors. Confining the map to the plane rules out the three-dimensional solution of forming a tetrahedral skeleton.
3. Rearrange the 12 matches to spell what matches are made of.

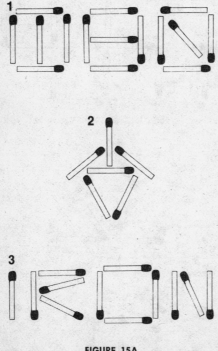

FIGURE 15A
Seven match puzzles

4. Change the positions of two matches to reduce the number of unit squares from five to four. "Loose ends"—matches not used as sides of unit squares—are not allowed. An amusing feature of this classic is that, even if someone solves it, you can set up the pattern again in mirror-reflected form or upside down (or both) and the solution will be as difficult as before.

5. It is easy to see how to remove four matches and leave two equilateral triangles, or to remove three matches and leave two equilateral triangles, but can you remove just *two* matches and leave two equilateral triangles? There must be no "loose ends."

FIGURE 15B

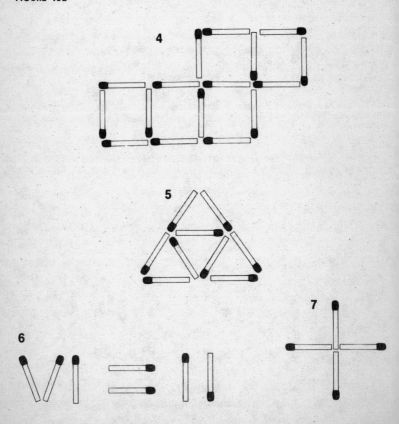

6. Move one match to produce a valid equation. Crossing the equality sign with a match to make it a not-equal sign is ruled out.

7. Move one match to make a square. (The old joke solution of sliding the top match up a trifle to form a square hole at the center is not permitted; here the solution is a different kind of joke.)

ADDENDUM

THE TWO MATCH GAMES are described as played with matches of different colored heads. If you can find a folder of black matches, then playing with black and white matches is even better, And of course both games can be played on paper by making an array of dots to be joined by lines of two different colors.

Nievergelt pointed out that David Silverman's proof of second-person win, on Connecto fields, does not apply when the game is played on other regular grids. For example, on a triangular lattice the first player can win by completing a unit triangle on or before his or her seventh move.

Nievergelt finds Connecto challenging on a variety of other grids, and wonders who has the win if the game is played on a cubical lattice. "It would be interesting," he writes, "if somebody found conditions of a graph-theoretical nature that would classify regular, infinite graphs as to whether the first player can force a circuit or not."

ANSWERS

DAVID SILVERMAN's puzzle is answered by observing that any player who wins the game of Connecto obviously must have

FIGURE 16
Solutions to match puzzles

two matches forming the letter L in the boundary of his region. The second player can prevent the first player from winning, on a board of any size, simply by playing to prevent his opponent from forming an L. If the first player plays the vertical bar of a possible L, the second player plays the horizontal bar. If the first player forms the horizontal part of a possible L, the second player forms the vertical part. This guarantees at least a draw for the second player.

Answers to the seven match puzzles are given in Figure 16. Several readers found an alternate solution to the sixth puzzle. The VI on the left is changed to XI, the Roman equivalent of the Arabic 11 on the other side.

CHAPTER 3

Spheres and Hyperspheres

"Mommy, Mommy, why do I always go 'round in circles?"
"Shut up or I'll nail your other foot to the floor."
—Children's sick joke, circa 1955

A CIRCLE is the locus of all points on the plane at a given distance from a fixed point on the plane. Let's extend this to Euclidean spaces of all dimensions and call the general n-sphere the locus of all points in n-space at a given distance from a fixed point in n-space. In a space of one dimension (a line) the 1-sphere consists of two points at a given distance on each side of a center point. The 2-sphere is the circle, the 3-sphere is what is commonly called a sphere. Beyond that are the hyperspheres of 4, 5, 6 . . . dimensions.

Imagine a rod of unit length with one end attached to a fixed point. If the rod is allowed to rotate only on a plane, its free end will trace a unit circle. If the rod is allowed to rotate in 3-space, the free end traces a unit sphere. Assume now that space has a fourth coordinate, at right angles to the other three, and that the rod is allowed to rotate in 4-space. The free end then generates a unit 4-sphere. Hyperspheres are impossible to visualize; nevertheless, their properties can be studied by a simple extension of analytic geometry to more than three coordinates.

A circle's Cartesian formula is $a^2 + b^2 = r^2$, where r is the radius. The sphere's formula is $a^2 + b^2 + c^2 = r^2$. The 4-sphere's formula is $a^2 + b^2 + c^2 + d^2 = r^2$, and so on up the ladder of Euclidean hyperspaces.

The "surface" of an n-sphere has a dimensionality of $n - 1$. A circle's "surface" is a line of one dimension, a sphere's surface is two-dimensional, and a 4-sphere's surface is three-dimensional. Is it possible that 3-space is actually the hypersurface of a vast 4-sphere? Could such forces as gravity and electromagnetism be transmitted by the vibrations of such a hypersurface? Many late-19th-century mathematicians and physicists, both eccentric and orthodox, took such suggestions seriously. Einstein himself proposed the surface of a 4-sphere as a model of the cosmos, unbounded and yet finite. Just as Flatlanders on a sphere could travel the straightest possible line in any direction and eventually return to their starting point, so (Einstein suggested) if a spaceship left the earth and traveled far enough in any one direction, it would eventually return to the earth. If a Flatlander started to paint the surface of the sphere on which he lived, extending the paint outward in ever widening circles, he would reach a halfway point at which the circles would begin to diminish, with himself on the *inside*, and eventually he would paint himself into a spot. Similarly, in Einstein's cosmos, if terrestrial astronauts began to map the universe in ever-expanding spheres, they would eventually map themselves into a small globular space on the opposite side of the hypersphere.

Many other properties of hyperspheres are just what one would expect by analogy with lower-order spheres. A circle rotates around a central point, a sphere rotates around a central line, a 4-sphere rotates around a central *plane*. In general the axis of a rotating n-sphere is a space of $n - 2$. (The 4-sphere is capable, however, of a peculiar double rotation that has no analogue in 2- or 3-space: it can spin simultaneously around two fixed planes that are perpendicular to each other.) The projection of a circle on a line is a line segment, but every point on the segment, with the exception of its end points, corresponds to two points on the circle. Project a sphere on a plane and you get a disk, with every point inside the circumference corresponding to two points on the sphere's surface. Project a 4-sphere on our 3-space and you get a solid ball with every internal point corresponding to two points on the 4-sphere's hypersurface. This too generalizes up the ladder of spaces.

The same is true of cross sections. Cut a circle with a line and the cross section is a 1-sphere, or a pair of points. Slice a sphere with a plane and the cross section is a circle. Slice a 4-sphere with a 3-space hyperplane and the cross section is a 3-sphere. (You can't divide a 4-sphere into two pieces with a 2-plane. A hyperapple, sliced down the middle by a 2-plane, remains in one piece.) Imagine a 4-sphere moving slowly through our space. We see it first as a point and then as a tiny sphere that slowly grows in size to its maximum cross section, then slowly diminishes and disappears.

A sphere of any dimension, made of sufficiently flexible material, can be turned inside out through the next-highest space. Just as we can twist a thin rubber ring until the outside rim becomes the inside, so a hypercreature could seize one of our tennis balls and turn it inside out through his space. He could do this all at once or he could start at one spot on the ball, turn a tiny portion first, then gradually enlarge it until the entire ball had its inside outside.

One of the most elegant of the formulas that generalize easily to spheres of all dimensions is the formula for the radii of the maximum number of mutually touching n-spheres. On the plane, no more than four circles can be placed so that each circle touches all the others, with every pair touching at a different point. There are two possible situations (aside from degenerate cases in which one circle has an infinite radius and so becomes a straight line): either three circles surround a smaller one [*Figure 17, left*] or three circles are inside a larger one [*Figure 17, right*]. Frederick Soddy, the British chemist who received a Nobel prize in 1921 for his discovery of isotopes, put

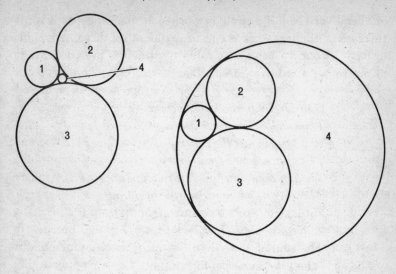

FIGURE 17
Find the radius of the fourth circle.

it this way in the first stanza of *The Kiss Precise*, a poem that appeared in *Nature* (Vol. 137, June 20, 1936, page 1021):

> *For pairs of lips to kiss maybe*
> *Involves no trigonometry.*
> *'Tis not so when four circles kiss*
> *Each one the other three.*
> *To bring this off the four must be*
> *As three in one or one in three.*
> *If one in three, beyond a doubt*
> *Each gets three kisses from without.*
> *If three in one, then is that one*
> *Thrice kissed internally.*

Soddy's next stanza gives the simple formula. His term "bend" is what is usually called the circle's curvature, the reciprocal of the radius. (Thus a circle of radius 4 has a curvature or "bend" of 1/4.) If a circle is touched on the inside, as it is in the case of the large circle enclosing the other three, it is said to have a concave bend, the value of which is preceded by a minus sign. As Soddy phrased all this:

> *Four circles to the kissing come.*
> *The smaller are the benter.*
> *The bend is just the inverse of*
> *The distance from the center.*
> *Though their intrigue left Euclid dumb*
> *There's now no need for rule of thumb.*
> *Since zero bend's a dead straight line*
> *And concave bends have minus sign,*
> The sum of the squares of all four bends
> Is half the square of their sum.

Letting a, b, c, d stand for the four reciprocals, Soddy's formula is $2(a^2 + b^2 + c^2 + d^2) = (a + b + c + d)^2$. The reader should have little difficulty computing the radii of the fourth kissing circle in each illustration. In the poem's third and last stanza this formula is extended to five mutually kissing spheres:

> *To spy out spherical affairs*
> *An oscular surveyor*
> *Might find the task laborious,*
> *The sphere is much the gayer,*
> *And now besides the pair of pairs*
> *A fifth sphere in the kissing shares.*
> *Yet, signs and zero as before,*
> *For each to kiss the other four*
> The square of the sum of all five bends
> Is thrice the sum of their squares.

The editors of *Nature* reported in the issue for January 9, 1937 (Vol. 139, page 62), that they had received several fourth stanzas generalizing Soddy's formula to *n*-space, but they published only the following, by Thorold Gosset, an English barrister and amateur mathematician.

> *And let us not confine our cares*
> *To simple circles, planes and spheres,*
> *But rise to hyper flats and bends*
> *Where kissing multiple appears.*
> *In n-ic space the kissing pairs*
> *Are hyperspheres, and Truth declares—*
> *As n + 2 such osculate*
> *Each with an n + 1-fold mate.*
> The square of the sum of all the bends
> Is *n* times the sum of their squares.

In simple prose, for *n*-space the maximum number of mutually touching spheres is *n* + 2, and *n* times the sum of the squares of all bends is equal to the square of the sum of all bends. It later developed that the formula for four kissing circles had been known to René Descartes, but Soddy rediscovered it and seems to have been the first to extend it to spheres.

Note that the general formula even applies to the three mutually touching two-point "spheres" of 1-space: two touching line segments "inside" a third segment that is simply the sum of the other two. The formula is a great boon to recreational mathematicians. Puzzles about mutually kissing circles or spheres yield readily to it. Here is a pretty problem. Three mutually kissing spherical grapefruits, each with a radius of three inches, rest on a flat counter. A spherical orange is also on the counter under the three grapefruits and touching each of them. What is the radius of the orange?

Problems about the packing of unit spheres do not generalize easily as one goes up the dimensional ladder; indeed, they become increasingly difficult. Consider, for instance, the problem of determining the largest number of unit spheres that can

FIGURE 18

Six unit circles touch a seventh.

touch a unit sphere. For circles the number is six [*see Figure 18*]. For spheres it is 12, but this was not proved until 1874. The difficulty lies in the fact that when 12 spheres are arranged around a 13th, with their centers at the corners of an imaginary icosahedron [*see Figure 19*], there is space between every pair. The waste space is slightly more than needed to accommodate a 13th sphere if only the 12 could be shifted around and properly packed. If the reader will coat 14 ping-pong balls with rubber cement, he will find it easy to stick 12 around one of them, and it will not be at all clear whether or not the 13th can be added without undue distortions. An equivalent question (can the reader see why?) is: Can 13 paper circles, each covering a 60-degree arc of a great circle on a sphere, be pasted on that sphere without overlapping?

H. S. M. Coxeter, writing on "The Problem of Packing a Number of Equal Nonoverlapping Circles on a Sphere" (in *Transactions of the New York Academy of Sciences*, Vol. 24,

FIGURE 19
Twelve unit spheres touch a thirteenth.

January 1962, pages 320–31), tells the story of what may be the first recorded discussion of the problem of the 13 spheres. David Gregory, an Oxford astronomer and friend of Isaac Newton, recorded in his notebook in 1694 that he and Newton had argued about just this question. They had been discussing how stars of various magnitudes are distributed in the sky and this had led to the question of whether or not one unit sphere could touch 13 others. Gregory believed they could. Newton disagreed. As Coxeter writes, "180 years were to elapse before R. Hoppe proved that Newton was right." Simpler proofs have since been published, the latest in 1956 by John Leech, a British mathematician.

How many unit hyperspheres in 4-space can touch a unit hypersphere? It is not yet known if the answer is 24, 25, or 26. Nor is it known for any higher space. For spaces 4 through 8 the densest possible packings are known only if the centers of the spheres form a regular lattice. These packings give lower bounds of 24, 40, 72, 126, and 240 for the number of unit spheres that can touch another. If we are not confined to regular lattice packings, the conjectured upper bounds are 26, 48, 85, 146, and 244. For spaces higher than 8, not even the densest regular packings are known. On the basis of nonlattice packing reported by Leech and N. J. A. Sloane in 1970, it is possible for 306 equal spheres to touch another equal sphere in 9 dimensions, and 500 can touch another in 10 dimensions. (The upper bounds are, respectively, 401 and 648.)

Why the difficulty with 9-space? A consideration of some paradoxes involving hypercubes and hyperspheres may cast a bit of dim light on the curious turns that take place in 9-space. Into a unit square one can pack, from corner to diagonally opposite corner, a line with a length of $\sqrt{2}$. Into a unit cube one can similarly pack a line of $\sqrt{3}$. The distance between opposite corners of an n-cube is $\sqrt{n}$, and since square roots increase without limit, it follows that a rod of *any* size will pack into a unit n-cube if n is large enough. A fishing pole 10 feet long will fit diagonally in the one-foot 100-cube. This also applies to objects of higher dimension. A cube will accommodate a square larger than its square face. A 4-cube will take a 3-cube larger than its cubical hyperface. A 5-cube will take larger squares and cubes than any cube of lower dimension with an edge of the same length. An elephant or the entire Empire State Building will pack easily into an n-cube with edges the same length as those of a sugar cube if n is sufficiently large.

The situation with respect to an n-sphere is quite different. No matter how large n becomes, an n-sphere can never contain a rod longer than twice its radius. And something very queer happens to its n-volume as n increases. The area of the unit circle is, of course, π. The volume of the unit sphere is 4.1+. The unit 4-sphere's hypervolume is 4.9+. In 5-space the volume is still larger, 5.2+, then in 6-space it decreases to 5.1+, and thereafter steadily declines. Indeed, as n approaches infinity the hypervolume of a unit n-sphere approaches zero! This leads to many unearthly results. David Singmaster, writing "On Round Pegs in Square Holes and Square Pegs in Round Holes" (*Mathematics Magazine*, Vol. 37, November 1964, pages 335–37), decided that a round peg fits better in a square hole than vice versa because the ratio of the area of a circle to a circumscribing square ($\pi/4$) is larger than the ratio of a square inscribed in a circle ($2/\pi$). Similarly, one can show that a ball fits better in a cube than a cube fits in a ball, although the difference between ratios is a bit smaller. Singmaster found that the difference continues to decrease through 8-space and then reverses: in 9-space the ratio of n-ball to n-cube is *smaller* than the ratio of n-cube to n-ball. In other words, an n-ball fits better in an n-cube than an n-cube fits in an n-ball if and only if n is 8 or less.

The same 9-space turn occurs in an unpublished paradox discovered by Leo Moser. Four unit circles will pack into a square of side 4 [*see Figure 20*]. In the center we can fit a smaller circle of radius $\sqrt{2} - 1$. Similarly, eight unit spheres will pack into the corners of a cube of side 4 [*see Figure 21*]. The largest

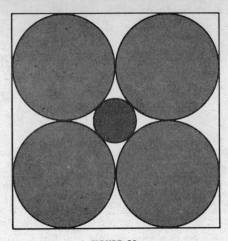

FIGURE 20
Four circles around one of radius $\sqrt{2} - 1$

sphere that will fit into the center has a radius of $\sqrt{3} - 1$. This generalizes in the obvious way: In a 4-cube of side 4 we can pack 16 unit 4-spheres and a central 4-sphere of radius $\sqrt{4} - 1$, which equals 1, so that the central sphere now is the same size as the others. In general, in the corners of an n-cube of side 4 we can pack 2^n unit n-spheres and presumably another sphere of radius $\sqrt{n} - 1$ will fit at the center. But see what happens when we come to 9-space: the central hypersphere has a radius of $\sqrt{9} - 1 = 2$, which is equal to half the hypercube's edge. The central sphere cannot be larger than this in any higher n-cube because it now fills the hypercube, touching the center of every hyperface, yet there is space at $2^9 = 512$ corners to take 512 unit 9-spheres!

A related unpublished paradox, also discovered by Moser, concerns n-dimensional chessboards. All the black squares of a chessboard are enclosed with circumscribed circles [*see Figure 22*]. Assume that each cell is of side 2 and area 4. Each circle has a radius of $\sqrt{2}$ and an area of 2π. The area in each white

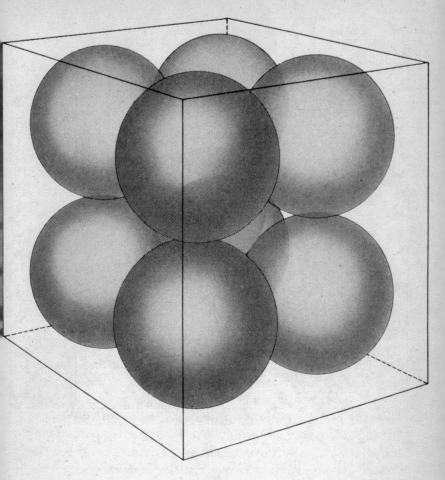

FIGURE 21

Eight unit spheres leave room for one with a radius of $\sqrt{3} - 1$.

cell that is left white (is not enclosed by a circle) is $8 - 2\pi = 1.71+$. In the analogous situation for a cubical chessboard, the black cubical cells of edge 2 are surrounded by spheres. The volume of each black cell is 8 and the volume of each sphere,

FIGURE 22
Leo Moser's hyperchessboard problem

which has a radius of $\sqrt{3}$, is $4\pi\sqrt{3}$, but the volume of the un-enclosed portion of each white cube is not so easy to calculate because the six surrounding spheres intersect one another.

Consider now the four-dimensional lattice of hypercubes of edge 2 with cells alternately colored as before so that each cell is surrounded by eight hypercubes of opposite color. Around each black hypercell is circumscribed a hypersphere. What is the hypervolume of the unenclosed portion within each white cell? The surprising answer can be determined quickly without knowing the formula for the volume of a hypersphere.

ANSWERS

THE FIRST problem was to determine the sizes of two circles, each of which touches three mutually tangent circles with radii of one, two, and three units. Using the formula given in the chapter,

$$2(1 + \frac{1}{4} + \frac{1}{9} + \frac{1}{x^2}) = (1 + \frac{1}{2} + \frac{1}{3} + \frac{1}{x^2})^2,$$

where x is the radius of the fourth circle, one obtains a value of $6/23$ for the radius of the smaller circle, 6 for the larger one.

The second problem concerned three grapefruits with three-inch radii and an orange, all resting on a counter and mutually touching. What size is the orange? The plane on which they rest is considered a fifth sphere of infinite radius that touches the other four. Since it has zero curvature it drops out of the formula relating the reciprocals of the radii of five mutually touching spheres. Letting x be the radius of the orange, we write the equation,

$$3(\frac{1}{3^2} + \frac{1}{3^2} + \frac{1}{3^2} + \frac{1}{x^2}) = (\frac{1}{3} + \frac{1}{3} + \frac{1}{3} + \frac{1}{x})^2,$$

which gives x a value of one inch.

The problem can, of course, be solved in other ways. When it appeared as problem 43 in the *Pi Mu Epsilon Journal*, November 1952, Leon Bankoff solved it this way, with R the radius of each large sphere and r the radius of the small sphere:

"The small sphere, radius r, touches the table at a point equidistant from the contacts of each of the large spheres with the table. Hence it lies on the circumcenter of an equilateral triangle, the side of which is $2R$. Then $(R + r)$ is the hypotenuse of a right triangle, the altitude of which is $(R - r)$ and the base of which is $2R\sqrt{3}/3$. So

$$(R + r)^2 = (R - r)^2 + 4R^2/3, \text{ or } r = R/3.\text{''}$$

The answer to Leo Moser's paradox of the hypercubic chessboard in four-dimensional space is that *no* portion of a white cell remains unenclosed by the hyperspheres surrounding each black cell. The radius of each hypersphere is $\sqrt{4}$, or 2. Since the hypercubic cells have edges of length 2, we see at once that each of the eight hyperspheres around a white cell will extend all the way to the center of that cell. The eight hyperspheres intersect one another, leaving no portion of the white cell unenclosed.

CHAPTER 4

Patterns of Induction

MANY GAMES AND PASTIMES have flimsy analogies with induction, that strange procedure by which scientists observe that some ostriches have long necks and conclude that all unobserved ostriches also have long necks. In poker and bridge, for instance, players use observational clues to frame probable hypotheses about an opponent's hand. A cryptographer guesses that a certain "pattern word," say BRBQFBQF, is NONSENSE, then tests this inductive conjecture by trying the letters elsewhere in the message. An old parlor entertainment involves passing a pair of scissors around and around a circle of players. As each person transfers the scissors he says "Crossed" or "Uncrossed." Those acquainted with the secret rule tell a player when he says the wrong word, and the joke continues until everyone has guessed the rule inductively. The scissors' blades are a red herring; a player should say "Crossed" if and only if his *legs* are crossed.

Familiar games such as Battleship and Jotto have slightly stronger analogies with scientific method, but the first full-fledged induction game was Eleusis, a card game invented by Robert Abbott and first explained in my *Scientific American* column for June 1959. (Fuller details are in *Abbott's New Card Games*, a Stein and Day hard-cover book in 1963 and a Funk & Wagnalls paperback in 1969.) Eleusis intrigued many mathematicians—notably Martin D. Kruskal of Princeton University, who worked out an excellent variant that he described in 1962 in a privately issued booklet, *Delphi—a Game of Inductive Reasoning*.

In Eleusis and Delphi a secret rule, specifying the order in which single cards may be played, corresponds to a law of nature. Players try to guess the rule inductively and then (like scientists) test their conjectures. In this chapter I shall explain a new type of induction game called Patterns, devised by Sidney Sackson and included in his delightful book *A Gamut of Games*.

Patterns is a pencil-and-paper game that can be played by any number of people, although preferably no more than six. It differs markedly from Eleusis and Delphi, but it shares with them such a striking similarity to scientific method that many thorny problems about induction, that have needled philosophers of science ever since David Hume showed induction has no logical justification, have pleasant analogues in the game.

Each player draws a square six-by-six grid on a sheet of paper. A player called the Designer (the role of Designer passes to another player with each new game) secretly fills in his 36 cells by drawing in each cell one of four different symbols. Sackson suggests the four shown in Figure 23, but any other four may be used. The Designer, who can be regarded as Nature, the Universe, or the Deity, is free to mark the cells as he likes; they may form a strong or a weakly ordered pattern, a partially ordered pattern, or no pattern at all. However (and here Sackson adopts the brilliant original idea of Abbott's), the method of scoring is such as to impel the Designer to create a pattern, or a regularity of nature, that is easy to discover for at

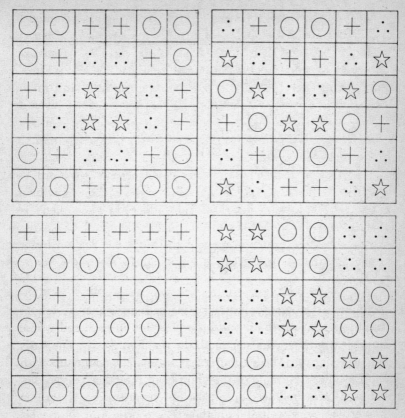

FIGURE 23

Patterns for Sidney Sackson's induction game,
all showing forms of symmetry

least one player and yet difficult enough to be missed by at least one other player.

Four typical patterns given in Sackson's book are arranged roughly in order of difficulty [*see Figure 23*]. All have some type of visual symmetry, but nonsymmetrical forms of order can be used if the players are mathematically sophisticated. For

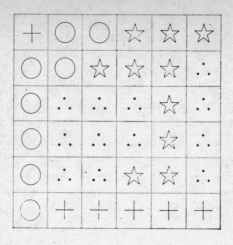

FIGURE 24
How is this pattern ordered?

example, a Designer might take the cells in sequence, left to right and top to bottom, putting a plus sign in each cell whose number is prime and a star in all the remaining cells. The basis for ordering the Master Pattern is intimately bound up with the Designer's estimate of the abilities of the other players because, as we shall see, he makes his highest score when one player does very well and another very poorly. Can the reader discern the simple basis for the nonsymmetrical ordering shown in Figure 24?

The Designer puts his sheet face down on the table. Any player may now make inquiries by drawing on his own grid a small slant line in the lower left-hand corner of any cell about which he seeks information. His sheet is passed face down to the Designer, who must enter the correct symbol in each cell in question. There are no turns. A player may ask for information whenever he wants, and there is no limit to the number of cells about which he may inquire. Each request represents an observation of nature—or an experiment, which is simply a controlled way of making special observations; cells filled in by the Designer correspond to the results of such observations. A player could ask for information about all 36 cells and obtain the entire pattern at once, but this is not to his advantage because, as we shall learn, it would give him a score of zero.

When a player believes he has guessed the Master Pattern, he draws symbols in all his untested cells. To make it easy to identify these inductions, guessed symbols are enclosed in parentheses. If a player decides he cannot guess the pattern, he may drop out of the game with a zero score. This is sometimes advisable because it prevents him from making a minus score and also because it inflicts a penalty on the Designer.

After all players have either filled in all 36 cells or dropped out of the game, the Designer turns his Master Pattern face up. Each player checks his guesses against the Master Pattern, scoring +1 for every correct symbol, −1 for every incorrect symbol. The sum is his final score. If he made a small number of inquiries and correctly guessed all or most of the entire pattern, his score will be high. If he has more wrong than right guesses, his score is negative. High scorers are the brilliant (or sometimes lucky) scientists; poor scorers are the mediocre, impulsive (or sometimes unlucky) scientists who rush poorly confirmed theories into print. Dropouts correspond to the mediocre or overcautious scientists who prefer not to risk framing any conjecture at all.

The Designer's score is twice the difference between the best and the worst scores of the others. His score is reduced if there are dropouts. Five points are subtracted for one dropout, 10 for each additional dropout. Sackson gives the following examples of games with a Designer (*D*) and players *A, B, C:*

If *A* scores 18, *B* scores 15, and *C* scores 14, *D*'s score is 8, or twice the difference between 18 and 14.

If *A* scores 18, *B* scores 15, and *C* scores −2, *D*'s score is 40, or twice the difference between 18 and −2.

If *A* scores 12, *B* scores 7, and *C* drops out with a score of 0,

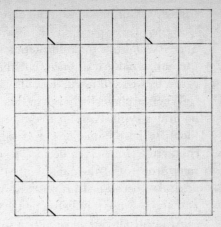

FIGURE 25
*Three stages in probing
for the Master Pattern*

D's score is 19, or twice the difference between 12 and 0, with five points deducted for the single dropout.

If *A* scores 12 and *B* and *C* both give up, *D* scores 9. This is twice the difference between 12 and 0, with five points deducted for the first dropout, 10 for the second.

If all three players drop out, *D*'s score is −25. His basic score is 0, with 25 points subtracted for the three dropouts.

An actual game played by Sackson suggests how a good player reasons [*see Figure 25*]. The five initial inquiries probe the grid for evidence of symmetry [*left*]. The sheet is returned with the five symbols filled in [*middle*]. A series of additional inquiries brings more information [*right*]. It looks as if the pattern is symmetrical around the diagonal axis from top left to bottom right. Since no stars have appeared, Sackson induces that they are absent from the pattern.

Now comes that crucial moment, so little understood, for the intuitive hunch or the enlightened guess, the step that symbolizes the framing of a hypothesis by an informed, creative scientist. Sackson guesses that the top left-hand corner cell contains a circle, that the three cells flanking it all have plus marks, and

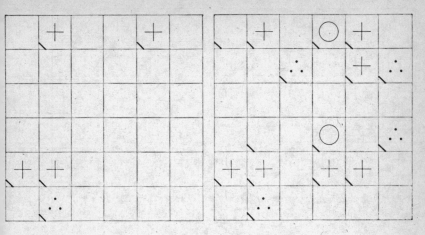

that, continuing down the diagonal, the pluses are flanked by three-spot symbols, the pattern repeating itself with larger borders of the same three symbols in the same order. To test this conjecture with as few new inquiries as possible, Sackson asks for information on only two more cells, the two cells shown empty but with slant lines on the grid at the right in Figure 25.

If those cells do not contain circles, his conjecture is false. As the philosopher Karl Popper maintains, the "strongest" conjecture is the one that is easiest to falsify, and Popper considers this the equivalent of the "simplest" conjecture. In Sackson's game the strongest (and simplest) conjecture is that every cell contains the same symbol, say a star. It is strong because a single inquiry about *any* cell, answered by anything but a star, falsifies it. The weakest conjecture is that each cell contains one of the four symbols. Such a hypothesis can be completely confirmed. Since no inquiry can falsify it, however, it is a true but useless hypothesis, empty of all empirical content because it tells one nothing about the Master Pattern.

The circles turn out to be where Sackson expected them. This increases what the philosopher Rudolf Carnap calls the "degree

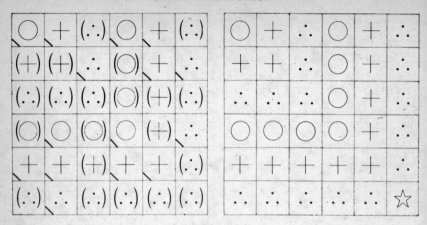

FIGURE 26

Player's grid (left) compared with Master Pattern (right)

of confirmation" of Sackson's hypothesis in relation to the total evidence he has bearing on it. Sackson decides to take the inductive plunge and "publish" his conjecture. He fills in the empty cells of his grid. When his pattern is compared with the Master Pattern [*see Figure 26*], a count of the guessed symbols (in parentheses) shows that Sackson has 20 right and one wrong, for a score of 19.

The single star Sackson missed is unexpected, but it is typical of the surprises Nature often springs. Science is a complicated game in which the universe seems to possess an uncanny kind of order, an order that it is possible for humans to discover in part, but not easily. The more one studies the history of the game of science, the more one has the eerie feeling that the universe is trying to maximize its score. A splendid recent example is the independent discovery by Murray Gell-Mann and Yuval Ne'eman of the "eightfold way." This is a symmetry pattern, defined by a continuous group structure, into which all the elementary particles seem to fit. As soon as enough information had accumulated the pattern was simple enough to be spotted by two physicists, and yet it remained complicated enough to be missed by all the other players.

Sackson, the inventor of Patterns, is a professional engineer who worked on steel bridges and buildings. Collecting, studying, and inventing games has been his lifelong avocation. He owns the largest private collection of modern proprietary games, books on games, and notes obtained by painstaking research in the world's great libraries and museums. He has invented hundreds of games. The first, he discloses in his book, was invented when he was in the first grade; it had to do with circling words on a page and joining them in chains. The first board game he owned was Uncle Wiggily, a track game that is still on the market. He immediately modified it by altering its rules and substituting toy soldiers for rabbits to make it a war game.

Almost all of Sackson's marketed games emphasize intellectual skill rather than luck. A game called Acquire, based on the theme of investing in hotel chains, has been his best-selling item. His other commercial games include The Case of the Elusive Assassin (a logic game based on Venn diagrams), Focus, Bazaar, Tam-Bit, Take Five, Odd or Even, Tempo, Interplay, and two card games, Venture and Monad.

A Gamut of Games is unique in that almost every one of its 38 games will be completely unfamiliar to any reader. All can be played with equipment that is easily acquired or constructed: cards, dice, dominoes, counters, and checkerboards. Twenty-two are Sackson originals. The others are either creations of Sackson's game-inventor friends or old, forgotten games that deserve revival. No two readers will, of course, have the same reaction to every game. I particularly like Knight Chase, played on a chessboard with one black and one white knight and 30 small counters. It is the invention of Alexander Randolph, Czechoslovakian born but now living in Venice, who has several excellent games on the U.S. market: Oh-Wah-Ree (based on the African game of mancala), Twixt, and Breakthru. Another mathematically appealing game (which Sackson found in an 1890 book) is Plank, a version of ticktacktoe played with 12 tricolored cardboard strips. The reader will find a valuable bonus in the book's final section: brief reviews of more than 200 of the best adult games on sale in this country.

Sackson's informal text is interspersed with personal anecdotes and snippets of surprising historical data. Until I read his book I did not know that the 17th-century poet Sir John Suckling invented cribbage, or that Monopoly, the most successful of all proprietary board games, is derived from The Landlord's Game, which was patented in 1904 by one Lizzie J. Magie and was intended to teach Henry George's single-tax theory. Sackson reproduces the patent drawing of the Magie board; the similarity to Monopoly is obvious.

Marketed board games, Sackson reminds us, tend to reflect major events and interests of the time. Although he does not mention it, an ironic example of this is The Money Game, a card game invented by Sir Norman Angell, who received the Nobel peace prize for 1933. The special cards and miniature money for this stock-market-speculation game were packaged with a 204-page explanatory book issued by E. P. Dutton, with puffs on the jacket by Walter Lippmann, John Dewey, and noted economists. Why is Angell's Money Game so grimly amusing? The year of its publication was 1929.

ADDENDUM

ROBERT ABBOTT has considerably modified his game of Eleusis to make it more exciting in actual play. For the rules of "New Eleusis," see my *Scientific American* column for October 1977.

Sidney Sackson retired in 1970 from his engineering chores to devote full time to inventing games and to writing. His *Gamut of Games* is still in print in hard cover (Castle Books), and currently on sale are four paperbacks he has done for Pantheon, a division of Random House: *Beyond Tic Tac Toe* (1975), *Beyond Solitaire* (1976), *Beyond Words* (1977), and

Beyond Competition (1977). All four contain tearout sheets for playing novel pencil-and-paper games. Sackson continues to review new games in his regular column in the war-gaming bimonthly *Strategy and Tactics*, and to contribute to the British magazine *Games and Puzzles* and to the new U.S. periodical *Games*.

More than two dozen of Sackson's original board games have been marketed in this country, of which the best known are his 3M games: Acquire, Bazaar, Executive Decision, Venture, Monad, and Sleuth. His game Focus is discussed in Chapter 5 of my *Sixth Book of Mathematical Games from Scientific American*.

Attempts to mechanize the process of induction by computer programs continues to be the object of much current research, and the topic of a growing body of literature. Several computer scientists experimented with programs for playing Sackson's game of Patterns. One such program is discussed in detail in Edward Thomas Purcell's *A Game-Playing Procedure for a Game of Induction*. It was Purcell's 1973 thesis for a master's degree in computer science at the University of California in Los Angeles.

ANSWERS

THE PROBLEM was to determine how a certain pattern for Sackson's induction game is ordered. The answer: Starting at the upper left-hand cell and spiraling clockwise to the center, there is first one symbol, then two symbols, then three, then four, then the same order of symbols is repeated in sets of five, six, seven, and eight.

CHAPTER 5

Elegant Triangles

ONE MIGHT SUPPOSE that the humble triangle was so thoroughly investigated by ancient Greek geometers that not much significant knowledge of the polygon with the fewest sides and angles could be added in later centuries. This is far from true. The number of theorems about triangles is infinite, of course, but beyond a certain point they become so complex and sterile that no one can call them elegant. George Polya once defined a geometric theorem's degree of elegance as "directly proportional to the number of ideas you see in it and inversely proportional to the effort it takes to see them." Many elegant triangle discoveries have been made in recent centuries that are both beautiful and important but that the reader is unlikely to have come across in elementary plane geometry courses. In this chapter we shall consider only a minute sample of such theorems, emphasizing those that have suggested puzzle problems.

"Ferst," as James Joyce says in the mathematical section of *Finnegans Wake*, "construct ann aquilittoral dryankle Probe loom!" We begin with a triangle, *ABC*, of any shape [*see Figure 27*]. On each side an equilateral triangle is drawn outward [*top left*] or inward [*top right*]. In both cases, when the centers

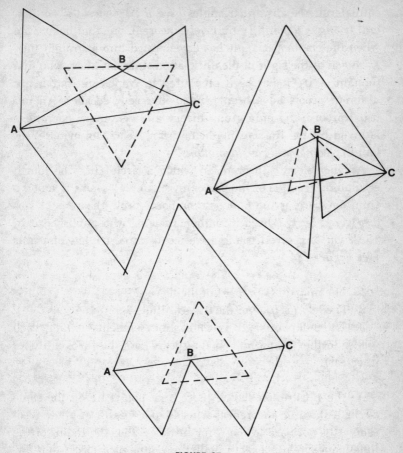

FIGURE 27

*Joining the centers of three equilateral triangles
creates a fourth one (dotted).*

(the intersections of two altitudes) of the three new triangles
are joined by straight [*dotted*] lines, we find we have constructed a fourth equilateral triangle. (The theorem is sometimes given in terms of constructing three isosceles triangles
with 30-degree base angles, then joining their apexes, but since
these apexes coincide with the centers of equilateral triangles,
the two theorems are identical.) If the initial triangle is itself

equilateral, the inward triangles give a "degenerate" equilateral triangle, a point. It is a lovely theorem, one that holds even when the original triangle has degenerated into a straight line, as shown at the right in the illustration. I do not know who first thought of it—it has been attributed to Napoleon—but many different proofs have been printed in recent decades. An unusual proof using only group theory and symmetry operations is given by the Russian mathematician Isaac Moisevitch Yaglom in *Geometric Transformations*.

Another elegant theorem, in which a circle (like the fourth equilateral triangle of the preceding example) seems to emerge from nowhere, is the famous nine-point-circle theorem. It was discovered by two French mathematicians, who published it in 1821. On any given triangle we locate three triplets of points [*see Figure 28*]:

1. The midpoints (a, b, c) of the three sides.
2. The feet (p, q, r) of the three altitudes.
3. The midpoints (x, y, z) of line segments joining each corner to the "orthocenter" (the spot where the three altitudes intersect).

As the illustration shows, those nine points lie on the same circle, a startling theorem that leads to a wealth of other theorems. It is not hard to show, for instance, that the radius of the nine-point circle is exactly half the radius of a circle that circumscribes the original triangle. The fact that the three altitudes of any triangle are concurrent (intersect at the same point) is interesting in itself. It is not in Euclid. Although Archimedes implies it, Proclus, a fifth-century philosopher and geometer, seems to have been the first to state it explicitly.

Three lines joining each midpoint of a side to the opposite vertex are called the triangle's medians [*see Figure 29*]. They too are always concurrent, intersecting at what is known as the triangle's centroid. The centroid trisects each median and the three medians carve the triangle into six smaller triangles of

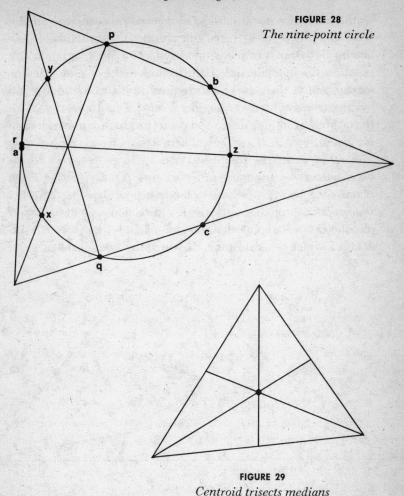

FIGURE 28
The nine-point circle

FIGURE 29
Centroid trisects medians

equal area. Moreover, the centroid is the triangle's center of gravity, another fact known to Archimedes. Your high school geometry teacher may have demonstrated this by cutting a scalene triangle from cardboard, drawing its medians to find the centroid, then balancing the triangle on a pencil by putting the centroid on the pencil's point.

The median is a special case of a more general line called a "cevian" (after a 17th-century Italian mathematician, Giovanni Ceva). A cevian is a line from a triangle's vertex to any point on the opposite side. If instead of midpoints we take trisection points, three cevians drawn as shown in Figure 30 will cut the triangle into seven regions, each a multiple of 1/21 of the original triangle's area. The central triangle, shown shaded, has an area of 3/21, or 1/7. There are many clever ways to prove this, as well as the results of a more general case where each side of the triangle is divided into n equal parts. If the cevians are drawn as before, to the first point from each vertex in a clockwise (or counterclockwise) direction around the triangle, the central triangle has an area of $(n-2)^2/(n^2-n+1)$. A still broader generalization, in which the sides of the original

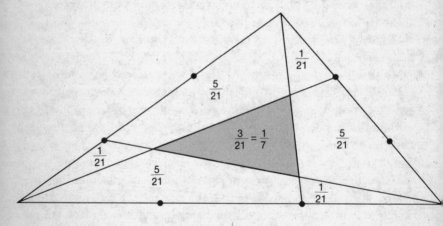

FIGURE 30
Trisecting cevians

triangle may vary independently in their number of equal parts, is discussed by H. S. M. Coxeter in his *Introduction to Geometry*. A formula going back to 1896 is given, and Coxeter shows how easily it can be obtained by embedding the triangle within a regular lattice of points.

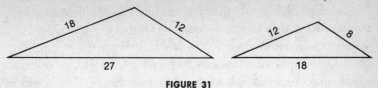

FIGURE 31
Smallest "5-con" triangle pair

Every triangle has three sides and three angles. Euclid proved three cases in which two triangles are congruent if only three of the six elements are equal (for example, two sides and their included angle). Is it possible for two triangles to have five of the six elements identical and yet *not* be congruent? It seems impossible, but there is an infinite set of such "5-con" triangles, as they have been called by Richard G. Pawley. Two 5-con triangles are congruent if three sides are equal, and therefore the only situation that permits noncongruence is the one in which two sides and three angles are equal. The smallest example of such a pair with integral sides is shown in Figure 31. Note that the equal sides of 12 and 18 are not corresponding sides. The triangles are necessarily similar, because corresponding angles are equal, but they are not congruent. The problem of finding all such pairs is intimately connected with the golden ratio.

There are many ancient formulas for finding a triangle's sides, angles or area, given certain facts about its altitudes, medians and so on. The expression $\sqrt{s(s-a)(s-b)(s-c)}$, where a, b, c are the sides of any triangle and s is half of the sum of the three sides, gives the triangle's area. This amazingly simple formula was first proved in the *Metrica* of Heron of Alexandria, who lived in the first or second century. The formula, Heron's chief claim to mathematical fame, is easily proved by trigonometry. Heron, or Hero as he is sometimes called, is best known today for his delightful treatises on Greek automata and hydraulic toys, such as the perplexing "Hero's fountain," in which a stream of water seems to defy gravity by spouting higher than its source.

A classic puzzle of unknown origin, the solution to which involves similar triangles, has become rather notorious because, as correspondent Dudley F. Church so aptly put it, "its charm lies in the apparent simplicity (at first glance) of its solution, which quickly evolves into an algebraic mess." The problem concerns two crossed ladders of unequal length. (The problem is trivial if the ladders are equal.) They lean against two buildings as shown in Figure 32. Given the lengths of the ladders and the height of their crossing point, what is the width of the space between the buildings? The three given values vary widely in published versions of the puzzle. Here we take a typical instance from William R. Ransom's *One Hundred Mathematical Curiosities*. The ladders, with lengths of 100 units (a) and 80 units (b), cross 10 units (c) above the ground. By considering similar triangles Ransom arrives at the formula, $k^4 - 2ck^3 + k^2(a^2 - b^2) - 2ck(a^2 - b^2) + c^2(a^2 - b^2) = 0$, which in this case becomes $k^4 - 20k^3 + 3,600k^2 - 72,000k + 360,000 = 0$.

This formidable equation is a quartic, best solved by Horner's method or some other method of successive approximations. The solution gives k a value of about 11.954, from which the width between buildings ($u + v$) is found to be 79.10+. There are many other approaches to the problem.

A difficult question now arises. Are there forms of this problem (assuming unequal ladders) in which all the labeled line segments in Figure 32 have integral lengths? As far as I know, this was first answered by Alfred A. Bennett in 1941 (see bibliography). His equations have since been rediscovered many times. The simplest solution (it minimizes both the height of the crossing and the width between buildings) is when the ladders are 119 and 70 units long, the crossing is 30 units above ground, and the width is 56 units. The number of solutions is infinite. There also are an infinite number of solutions in which the distance between the *tops* of the ladders is also an integer. (See Gerald J. Janusz's solution, cited in the bibliography.)

If we require only that the lengths of the ladders, the dis-

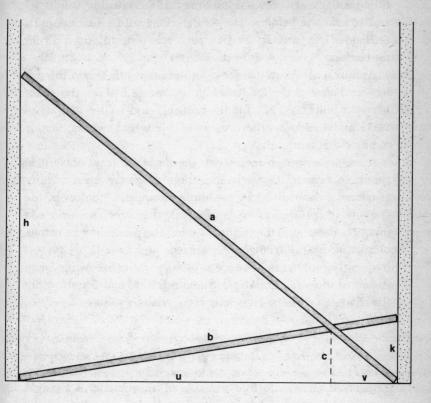

FIGURE 32
Crossed-ladders problem

tance between buildings, and the height of the crossing be integers, we can then seek answers that minimize particular
values. H. G. ApSimon sent the most complete analysis. The
solution that minimizes the space between buildings is 40 for
this variable, 38 for the height of crossing, 58 and 401 for the
ladders. The height of the crossing has a minimum of 14 when
the space between buildings is 112 and the ladders are 113 and
238. (Both these solutions had earlier been found by John W.
Harris.) The solution that minimizes the longest ladder length

is 63 for the space between buildings, 38 for crossing height, 87 and 105 for the ladders. The solution that minimizes the shortest ladder length is 40 for the space between buildings, 38 for the crossing, 58 and 401 for the ladders.

ApSimon also searched for a solution that minimizes the difference between the ladders. His best was 1,540 for the space between buildings, 272 for the crossing, and ladder lengths of 1,639 and 1,628—a difference of 11. He was, however, unable to prove this minimal.

If we are given no more than the distances from a point to the three vertexes of a triangle, there obviously is an infinity of triangles determined by the three distances. If, however, the triangle is required to be equilateral, the three distances can uniquely determine the triangle's side. The point may be inside, outside, or on the triangle. An ancient problem of this type is frequently sent to me by readers, usually in the following form. A point inside an equilateral triangle is 3, 4, and 5 units from the triangle's corners. How long is the triangle's side?

ANSWERS

THE PROBLEM was to find the side of an equilateral triangle containing a point p that is three, four, and five units from the triangle's corners. The following solution is from Charles W. Trigg, *Mathematical Quickies*. The broken lines of Figure 33 are constructed so that PCF is an equilateral triangle and AE is perpendicular to PC extended left to E. Angle $PCB = 60$ degrees minus angle $PCA =$ angle ACF. Triangles PCB and FCA are therefore congruent and $AF = BP = 5$. Because APF is a right triangle, angle $APE = 180 - 60 - 90 = 30$ degrees. From this we conclude that AE is 2 and EP is twice the square root of 3. This permits the equation

$$AC = \sqrt{2^2 + (3 + 2\sqrt{3})^2}$$
$$= \sqrt{25 + 12\sqrt{3}},$$

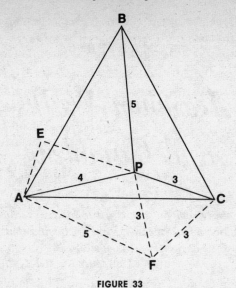

FIGURE 33
Solution to three-distances problem

which gives AC, a side of the original triangle, a value of 6.766+.

There is a beautifully symmetric equation for finding the side of an equilateral triangle when given the distances of a point from its three corners:

$$3(a^4 + b^4 + c^4 + d^4) = (a^2 + b^2 + c^2 + d^2)^2.$$

Any three variables can be taken for the three distances. Solving for the fourth then gives the triangle's side. The simplest solution in integers is 3, 5, 7, 8. The point is outside the triangle except when the side is 8, when it lies on a side of the triangle. W. H. Grindley, Jörg Waldvogel, and others sent proofs that in all three cases (point inside, outside, or on the triangle) there is an infinity of primitive (no common divisor) integral solutions. The simplest is 57, 65, and 73 for the distances, and 112 for the triangle's side. The next simpler: 73, 88, 95, and 147.

CHAPTER 6

Random Walks and Gambling

> *He calmly rode on, leaving it to his horse's discretion to go which way it pleased, firmly believing that in this consisted the very essence of adventures.*
> —DON QUIXOTE, *Vol. 1, Chapter 2*

THE COMPULSIVE DRIFTER who wanders aimlessly from town to town may indeed be neurotic, and yet even the sanest person needs moderate amounts of random behavior. One form of such behavior is traveling a random path. Surely the popularity of the great picaresque novels such as *Don Quixote* is due partly to the reader's vicarious pleasure in the unexpectedness of events that such haphazard paths provide.

Jorge Luis Borges, in his essay "A New Refutation of Time," describes a random walk through the streets of Barracas: "I tried to attain a maximum latitude of probabilities in order not to fatigue my expectation with the necessary foresight of any one of them." G. K. Chesterton's second honeymoon, as he describes it in his autobiography, was a random "journey into the void." He and his wife boarded a passing omnibus, left it when they came to a railway station, took the first train and at the end of the line left the train to stroll at random along country roads until they finally reached an inn, where they stayed.

Mathematicians insist on analyzing anything analyzable. The random walk is no exception and (mathematically speaking) is as adventurous as the wanderings of the man of La Mancha. Indeed, it is a major branch of the study of Markov chains, which in turn is one of the hottest aspects of modern probability theory because of its increasing application in science.

A Markov chain (named for the Russian mathematician A. A. Markov, who first investigated them) is a system of discrete "states" in which the transition from any state to any other is a fixed probability that is unaffected by the system's past history. One of the simplest examples of such a chain is the random walk along the line segment shown in Figure 34. Each interval on the line is a unit step. A man begins the walk at spot 0. He flips a coin to decide the direction of each step: heads he goes right, tails he goes left. In mathematical terminology his "transition probability" from one mark to the next is 1/2. Since he is just as likely to step to the left as to the right, the walk is called "symmetric." Vertical bars A and B, at -7 and $+10$, are "absorbing barriers." This means that if the man steps against either barrier, it "absorbs" him and the walk ends.

A novel feature of this walk is its isomorphism with an ancient betting problem called "gambler's ruin." Player A starts

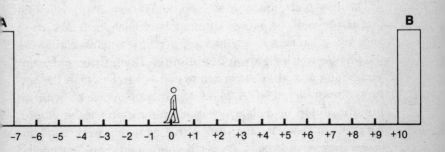

FIGURE 34
One-dimensional random walk with absorbing barriers

with $7, *B* with $10. They repeatedly flip a coin. For each head *B* gives *A* $1 and for each tail *A* gives *B* $1. The game ends when either player is "ruined," or runs out of money. It is easy to see the correspondence between this game's progress and the random walker's movements. At any moment *A*'s capital in dollars is represented by the walker's distance from barrier *A*, *B*'s capital by the walker's distance from barrier *B*. If the first two tosses are heads, the walker moves two steps to the right; in the betting interpretation, *A* has increased his capital from $7 to $9, whereas *B* has gone from $10 to $8. If the walker hits barrier *A*, it corresponds to *A*'s ruin. If he hits barrier *B*, it corresponds to *B*'s ruin.

All kinds of probability questions have identical answers in both interpretations. Some are easy to solve, some are extremely difficult. One of the easiest is: What is the probability of each player's winning? This is the same as asking for the probability that the walk will end at one barrier or the other. It is not hard to prove that the probability of each man's winning is given by his original capital divided by the total number of dollars held by both players. *A*'s probability of winning is 7/17, *B*'s is 10/17. In random-walk terms the probability that the walk ends at barrier *B* is 7/17, at barrier *A* 10/17 [for a simple proof of this, based on a stretched rubber band, see "Brownian Motion and Potential Theory," by Reuben Hersh and Richard J. Griego, in *Scientific American*, March 1969].

The two probabilities must add to 1 (certainty), meaning that if the walk or game continues long enough, it is sure to end. What happens if one barrier, say *B*, is removed, allowing the line to stretch rightward to infinity? Then, if the walk continues long enough, it is certain to end at barrier *A*. In the betting interpretation, if *A* plays against an opponent with an infinite supply of dollars, eventually *A* is sure to be ruined. This is bad news for the compulsive gambler: even if all his wagers are at fair odds, he is playing against an "opponent" (the gambling world) with virtually unlimited capital, making his eventual ruin almost certain.

Another type of easy calculation is the probability that the walker, starting at a certain spot, will reach another given spot (or return to his starting point) after a specified number of steps. Odd and even parity is involved, so that in half the cases the answer is 0 (impossible). For instance, the walker cannot go from 0 to an even-numbered spot in an odd number of steps or to any odd-numbered spot in an even number of steps. What is the probability that he will walk from 0 to +1 in exactly three steps? It is the same as the probability that three coin tosses will show, in any order, two heads and one tail. Since this happens three times in the eight equally probable outcomes, the answer is 3/8. (The situation can be complicated by replacing either or both absorbing barriers by a "reflecting barrier" halfway between two marks. When the walker hits such a barrier, he bounces back to the mark he has just left. In the betting interpretation this happens when a ruined gambler is given $1 so that he can stay in the game. If both barriers are reflecting, of course, the walk never ends.)

Another simple calculation, although one that is harder to prove, is the expected number of tosses before the walker, on the line with two absorbing barriers, is absorbed. "Expected number" is the average in the long run of repeated repetitions of the walk. The answer is the product of the distances of the two barriers from the starting spot. In this case, $7 \times 10 = 70$. The "typical" walk lasts for 70 steps; the typical game ends with one player ruined after 70 coin tosses. This is considerably longer than most people would guess. It means that in a fair betting game between two players, each starting with $100 and each making $1 bets, the average game will last for 10,000 bets. Even more counterintuitive: If one man starts with $1 and the other with $500, the average game will last for 500 bets. In the random walk, if the man begins one step from one barrier and 500 steps from the other, his average walk before being absorbed is 500 steps!

What is the expected number of steps until a walker first reaches a distance n from 0, the starting spot, assuming that neither barrier (if any) is closer to the start than n? This is easily seen to be a special case of the above problem. It is the same as asking for the expected number of steps until the walker is absorbed, when each barrier is a distance n from 0. It is simply $n \times n = n^2$. Thus if a walker takes n steps and finds himself at a maximum distance from 0, the expected distance is $\sqrt{n}$.

This is not the same as asking for the expected distance from 0 after n steps when the distance need not be maximum. In this case the formula is a bit trickier. For one step it is obviously 1, for two steps it is also 1 (the four equally possible distances are 0, 0, 2, 2). For three steps it is 1.5. As n approaches infinity the limit for the expected distance (which may be on either side of 0) is $\sqrt{2n/\pi}$, or about $.8\sqrt{n}$ for large n, as Frederick Mosteller and his coauthors point out in their book *Probability and Statistics*, page 14.

The hardest to believe of all aspects of the one-dimensional walk emerges when we consider a walk starting on a line with no barriers and ask how often the walker is likely to change sides. Because of the walk's symmetry one expects that in a long walk the man should spend about half of his time on each side of the starting spot. Exactly the opposite is true. Regardless of how long he walks, the most probable number of changes from one side to the other is 0, the next most probable is 1, followed by 2, 3, and so on!

William Feller, in a famous chapter on "Fluctuations in Coin Tossing and Random Walks" (in his classic *An Introduction to Probability Theory and Its Applications*, Vol. I, Chapter III), has this to say: "If a modern educator or psychologist were to describe the long-run case histories of individual coin-tossing games, he would classify the majority of coins as maladjusted. If many coins are tossed n times each, a surprisingly large proportion of them will leave one player in the lead almost all the time; and in very few cases will the lead change sides and fluctuate in the manner that is generally expected of a well-behaved coin." In a mere 20 tosses the probability that each player will lead 10 times is .06+, the least likely outcome. The probability that the loser will *never* be in the lead is .35+.

If a coin is tossed once a second for a year, Feller calculates, in one out of 20 repetitions of this experiment the winning player can be expected to lead for more than 364 days and 10 hours! "Few people will believe," he writes, "that a perfect coin will produce preposterous sequences in which no change of lead occurs for millions of trials in succession, and yet this is what a good coin will do rather regularly."

Figure 35 is a graph of a typical random walk along the infinite vertical line at the left, with time represented by movement to the right. Instead of flipping a coin or using a table of random numbers, the walk is based on the digits of pi to 100 decimals. (Since the decimals of pi have passed all randomness tests, they provide a convenient source of random digits.) Each even digit is a step up, each odd digit a step down. After 101 steps the walker has been above the line only 17 times, about 17 percent of the total. He has crossed the starting spot only once. The graph is also typical in showing how returns to 0 or close to 0 come in waves that tend to increase in length at a rate about equal to the square root of the time. Similar graphs based on simulations of 10,000 coin tosses appear in Feller's book.

We can complicate matters by allowing transition probabilities to vary from 1/2 and by allowing steps longer than one unit. Consider the following curious paradox first called to my

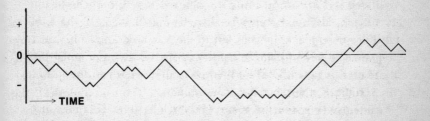

FIGURE 35

Symmetric random walk based on the first 101 digits of pi

attention (in betting terms) by Enn Norak, a Canadian mathematician. A walker starts 100 steps to the right of 0 on a line that has no barriers [*see Figure 36*]. Instead of a coin a packet of 10 playing cards—five red and five black—is used as a randomizer. The cards are shuffled and spread face down and any card is selected. After its color is noted it is discarded. If it is

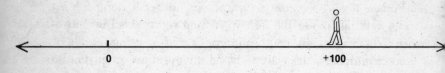

$$0 \qquad\qquad\qquad\qquad +100$$

FIGURE 36
Paradox based on a random walk along a line without barriers

red, the walker steps to the right; if black, he steps to the left. This continues until all 10 cards have been taken. (The transition probability varies with each step. It is 1/2 only when there is an equal mixture of red and black cards before the draw.) The walk differs also from walks discussed above in that before each card is noted the walker chooses the length (which need not be integral) of his next step.

Assume that the walker adopts the following halving strategy in choosing step lengths. After each card is noted he takes a step (left or right) equal to exactly half of his distance from 0. His first step is $100/2 = 50$ units. If the card is red, he goes to the 150 mark. His next step will then be $150/2 = 75$. If the first card drawn is black, he goes left to the 50 mark, and so his next step will be $50/2 = 25$. He continues in this manner until the tenth card is noted. Will he then be to the right or to the left of the 100 mark where he began the walk?

The answer is that he is sure to be to the left. This may not be very surprising, but it is surely astonishing that, regardless of the order in which the cards are drawn, he will end the walk at exactly the same spot! It is about 76 units left of where he started. The precise distance is given by the following formula,

$$a - \left[a \left(\frac{3}{4} \right)^n \right],$$

where a is the starting spot and n the number of red (or black) cards in the packet. When a is 100 and n is 5, as in the present example, the formula gives 76.26953125 as the distance he has moved to the left when the walk ends.

Let us translate this into Norak's betting game. A man starts with \$100. Wins and losses are decided by a shuffled packet of five red and five black cards, from which cards are drawn and discarded. (This is equivalent to flipping a coin 10 times, provided that the coin happens to show an equal number of heads and tails. Using cards guarantees this equality.) The man wins on red and loses on black. Each time he bets half of his capital. It is hard to believe, but at the end of every such game he will have lost exactly \$76.26953125. This amount increases as n increases. If n is 26, as it is if a standard deck of 52 cards is used, he will lose more than \$99.90. His loss, however, will always be less than \$100.

Instead of betting half of his capital each time, he can bet a fixed fraction. Let the fraction be $1/k$, where k is any positive real number. The smaller this fraction is, the less he will have lost by the end of the game; the larger the fraction, the more he will have lost. If it equals 1, he is certain to lose everything. In this more general case the amount lost is

$$a - \left[a \left(1 - \frac{1}{k^2} \right)^n \right].$$

The formula can be generalized further by allowing an unequal mixture of red and black cards, but this gets too complicated to explore here.

Now consider an amusing problem suggested by Norak and based on a variation of the game just described. It can be given as a random walk problem but I shall give only its betting equivalent. The game is the same as before except that the *opponent* of the man who starts with $100 is allowed to name the size of each bet. Call the opponent Smith and assume that he has enough capital to be able to pay any loss. A standard deck of 52 playing cards is used. Before each card is drawn and discarded, Smith bets exactly half the capital then owned by the *other* man, the player who begins with $100. After the last card is noted will Smith have lost or gained? In either case, is the loss or gain always the same and, if so, what is its formula? If you have understood the discussion to this point, you should be able to answer these questions almost immediately.

The staggering topic of random walks will be concluded in the next chapter with a consideration of some random walks on the plane and in space, and on lattices such as checkerboards and the edges of regular solids.

ANSWERS

THE BETTING PROBLEM was intended as a joke. If player A begins with a certain number of dollars and if, on each draw of a card from a packet with an equal number of red and black cards, A's opponent B is allowed to bet half of A's current capital, the game obviously is the same as the one explained earlier, in which A always bets half of his *own* capital. B, who now calls the bets, will win precisely what A would have lost in the earlier game. Therefore the formula given for the first game applies here. We were told that the loser starts with $100 and that a deck of 52 cards is used as a randomizer. The winner is sure to win exactly $100 - [100(3/4)^{26}]$ dollars, leaving the loser with less than a dime.

CHAPTER 7

Random Walks on the Plane and in Space

IN THE PREVIOUS CHAPTER I discussed the random walk of discrete steps along a line, with or without absorbing barriers, and pointed out its curious equivalence to various two-person betting games. In this chapter the random walk moves onto the plane and into space.

One type of two-dimensional random walk that has been much studied is one that goes from vertex to adjacent vertex on the infinite checkerboard lattice shown in Figure 37. Every step is one unit and the walk is "symmetric" in the sense that each of the four possible directions is picked with a probability of 1/4. The walk can be made finite by surrounding the walker with absorbing barriers, shown as dots; when he steps on one of these dots, he is "absorbed" and the walk ends. (The surrounding barriers need not form a neat square. They may form a boundary of any shape.) As in the analogous finite walk on the line, it is not hard to calculate the probability that a walk starting at any vertex inside the boundary will end at a specified barrier. One can also determine the expected number of steps (the average in the long run of repeated walks) that will be taken before a walk ends. The formulas involved in such calculations have many unexpected scientific applications, such as determining the voltages at interior parts of electrical networks.

When the walker is not trapped inside barriers but can es-

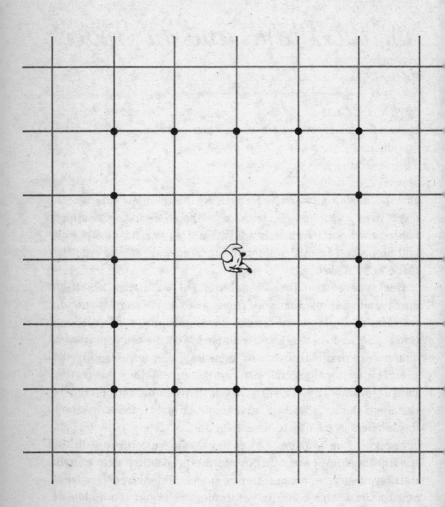

FIGURE 37

Random walk on a square lattice

cape to wander over a square lattice that covers the infinite plane, the situation grows more complicated and gives rise to many problems that are as yet unsolved. Some of the established theorems are deep and paradoxical. Consider a random walk on an infinite lattice with no barriers. If the walk continues an arbitrarily long time, the proportion of visits the walker makes to any specified corner approaches zero as a limit. On the other hand, if the walk continues long enough, the walker is certain to touch every vertex, including a return visit to his starting spot. As John G. Kemeny points out in "Random Walks," an excellent nontechnical article in *Enrichment Mathematics for High School*, this introduces a profound distinction between logical and practical possibility. It is *logically* possible that such a walker can travel forever without reaching a given corner. To the mathematician, however, it has a *practical* probability of zero even though the expected number of steps for reaching any specified corner is infinite. The distinction is often encountered where infinite sets are concerned. If a penny is flipped forever, for example, it is logically possible that heads and tails will forever alternate, although the practical probability that this will happen is zero.

Kemeny expresses it this way: If you stand at an intersection on the infinite lattice while a friend, starting at any other spot, wanders randomly over the lattice, he will be practically certain to meet you if you are able to wait an arbitrarily long time. The statement can be even stronger. After the first meeting the probability is again 1 that if your friend continues wandering, he will eventually return to you. In other words, it is practically certain that such a walker, given enough time, will visit every intersection an infinity of times!

Suppose two walkers move haphazardly over an infinite square lattice. Are they certain to meet? (If they begin an odd number of steps apart and step in unison, they can never meet at a corner, but they can bump into each other at the middle of a line segment.) Once more the answer is that they will meet infinitely often if they walk long enough. If three men step in unison and wander over the infinite lattice, and if each pair of their starting spots is separated by an even number of steps, all three are certain to meet at *some corner*. The probability of their meeting at a specified corner, however, drops to less than 1. For four or more walkers the probability that all four will meet *somewhere* also becomes less than 1.

The biggest surprise comes when we generalize to a space lattice. If such a lattice (it need not be cubical) is finite, a random walker is practically certain to reach any intersection in a finite time. As Kemeny puts it, if you are inside a large building with a complex network of corridors and stairways, you can be sure of reaching an exit in a finite time by walking randomly through the building. If the lattice is infinite, however, this is not the case. George Polya proved in 1921 that the probability is less than 1 that a random walker will reach any assigned corner on such a lattice even if he walks forever. In 1940 W. H. McCrea and F. J. W. Whipple showed that the probability of a walker's returning to his starting spot, after wandering an infinite time on an infinite cubical lattice, is about .35.

Turning from planar lattices to the plane itself, allowing the walker to step a unit distance in any randomly selected direction, the situation becomes more complicated in some ways and simpler in others. For example, the expected (average) distance of the walker from his starting spot, after n equal steps, is simply the length of the step times the square root of n. This was proved by Albert Einstein in a paper on molecular statistics published in 1905, the same year he published his celebrated first paper on relativity. (It was independently proved by Marian Smoluchowski. Readers will find a simple proof in George Gamow's *One Two Three . . . Infinity*.)

Discrete random walks in space obey the same square-root formula. As on the plane, the steps need not be uniform in length. The expected distance from the origin, after n steps, is the average length of a step times the square root of n. It is here that random walks become invaluable in the study of diffusion phenomena: the random movements of molecules in a liquid or gas, the diffusion of heat through metals, the spread of rumors, the spread of a disease, and so on. The drift of a flu epidemic is a blend of millions of random walks by microbes. There are applications in almost every science. The first major use of the Monte Carlo method—a way of using computers to simulate difficult probability problems—was in calculating the random walks of neutrons through various substances. In such diffusion phenomena, as well as in Brownian motion, the square-root formula must be modified by many other factors such as temperature, the viscosity of the embedding medium, and so on. Moreover, such motions are usually continuous, not discrete; they are called Markov "processes," as distinct from Markov "chains." The square-root formula provides only a first approximation for estimating expected distances. (For recent work in this field, beginning with Norbert Wiener's brilliant first paper on Brownian motion in 1920, see "Brownian Motion and Potential Theory," by Reuben Hersh and Richard J. Griego, in *Scientific American*, March 1969.)

The outward drift of a random walker from his starting spot, on the plane or in space, is not at a constant rate. If the walk itself is at a steady pace, the square root of the number of steps increases at a steadily decreasing rate. The longer the walk, the slower the drift. Gamow, in the book cited above, gives a dramatic illustration. A light quantum near the sun's center takes about 50 centuries to perform a "drunkard's walk" to the surface. Once free of the sun, it instantly sobers up and, if it is headed in the right direction, reaches the earth in about eight minutes.

Here is a simple question. Two men start at the same spot on the plane. One makes a random walk of 70 unit steps, then stops. The other stops after a random walk of 30 unit steps. What is the expected distance between them at the finish?

We turn now to a type of random walk that is different from any we have considered so far. Assume that a bug starts at cor-

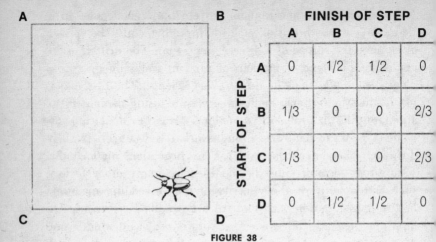

FIGURE 38

"Ergodic walk" on a square (left) and the matrix of its transition probabilities (right)

ner *A* of the square shown at the left of Figure 38 and crawls randomly along its edges. Instead of equalizing the "transition probabilities" from corner to corner, as in earlier examples, assume that at corners *B* and *C* the probability of the bug's heading toward *D* is twice the probability of its returning to *A*. At *A* and *D* the bug chooses between two paths with a probability of 1/2 for each, whereas at *B* and *C* it chooses the path to *D* with a probability of 2/3 and the path to *A* with a probability of 1/3. The network is finite but, since there are no absorbing barriers, the walk never ends. Such a walk is usually called an "ergodic walk." We should like to calculate what fraction of visits, in the long run, the bug makes to each corner.

One way to do it is to draw the "stochastic matrix" at the right in the illustration, which shows the transition probabilities from any corner to any other. Zeros on the matrix indicate transitions that cannot occur. Since every state of this ergodic Markov chain must lead to another, the sum of the probabilities on any horizontal row, called a "probability vector," must be 1.

The probability that the bug will visit a given corner is the same as the sum of the probabilities that it will move to that corner from an adjacent corner during its endless walk. For instance, the probability that it is at D is the probability that it will go to D from B added to the probability that it will go to D from C. (These are long-run probabilities, not the probabilities of going to D when the bug *starts* at B or C.) Let d be the probability that at any moment when the bug is at a corner it will be at corner D. Let a, b, c be the probabilities that it is at corners A, B, C. From the D column of the matrix we can see that the probability that, in the long run, the bug will be going from B to D is $b(2/3)$ and from C to D is $c(2/3)$. The long-run probability of being at D is the sum of these two probabilities, and therefore we can write the following equation: $d = b(2/3) + c(2/3)$. This simplifies to

$$d = 2b/3 + 2c/3.$$

The other three columns give us similar formulas for a, b, and c:

$$a = b/3 + c/3$$
$$b = a/2 + d/2$$
$$c = a/2 + d/2$$

When the bug is not on an edge, it is certain to be at a corner, and so we have a fifth equation:

$$a + b + c + d = 1.$$

A glance at the four preceding equations shows that $b = c$ and $d = 2a$, making it easy to solve the five simultaneous equations: $a = 1/6$, $b = 1/4$, $c = 1/4$, $d = 1/3$. The bug will spend 1/6 of its corner time at A, 1/4 at B, 1/4 at C, 1/3 at D. It will make twice as many visits to D as to A.

Readers might like to try the same technique on the cube analogue of this problem, given by Kemeny in the article men-

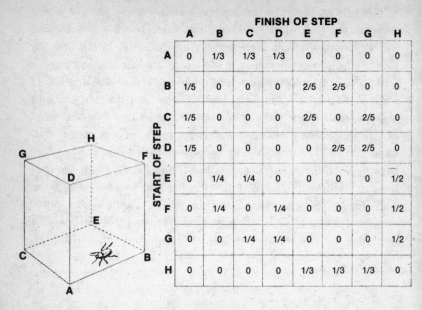

FIGURE 39

*Ergodic random walk on a cube (left) and the matrix of its
transition probabilities (right)*

tioned above. On the cube shown at the left in Figure 39, the
bug is twice as likely to step toward H as toward A. The stochastic matrix of transition probabilities is shown to the right
of the cube. The eight simultaneous equations derived from the
eight columns, together with the equality $a + b + c + d + e +
f + g + h = 1$, have a unique solution. The bug, performing its
perpetual ergodic walk, will make 3/54 of its corner visits to A,
5/54 to each of B, C, D, 8/54 to each of E, F, G, and 12/54 to
H. It will visit H four times as often as A.

If an ergodic walk of this type is symmetric, in the sense that
at each vertex the possible next steps are chosen with equal
probability, the fraction of visits spent at any two given corners
is proportional to the number of different ways to walk to those
two corners. For example, a cat performing a symmetric random ergodic walk along the edges of the Great Pyramid of

Egypt will visit the apex four times for every three times it visits a base corner because there are four ways to walk to the top corner and only three ways to reach each base corner. It is easy to draw the matrix and write the equations showing that the cat will make 1/4 of its visits to the apex and 3/16 to each base corner.

For another easy problem, assume that on the cube shown in Figure 39, with the same matrix, a fly starts a random walk at *A*. At the same time a spider begins a random walk at *H*. Both move at the same speed. What is the probability that they will meet at the middle of an edge after each has gone the minimum distance of 1½ edges?

Many other pleasant problems arise in connection with ergodic random walks along the edges of a cube and other regular solids. If a drunk bug starts at a corner of a cube and walks to the most distant corner, at each corner choosing one of the three paths with equal probability, its average walk will have a length of 10 edges. If the bug is only semidrunk, never going back along an edge just traversed but selecting the other two paths with equal probability, its average walk to the most distant corner is 6. In *both* cases the bug's average walk back to its starting corner is 8, the number of corners on a cube.

This is no coincidence. Thomas H. O'Beirne of Glasgow (whose term "semidrunk" I have borrowed) has shown in an unpublished proof that, on any regular network with every vertex topologically the same as every other, a random walk back to a starting vertex has an expected number of steps equal to the total number of vertices. It is true regardless of whether the walker chooses all paths at each vertex with equal probability or only those that exclude the path just traveled. A drunk or semidrunk bug stepping from corner to corner on a square will have an average walk back to the starting corner of four steps. The edges of all Platonic and Archimedean solids form regular space networks of the same kind. On a tetrahedron a drunk or semidrunk bug will traverse four edges in an average walk back to a starting corner; on a dodecahedron it will traverse 20 edges, and so on. Readers interested in how to set up equations for calculating average walks on such networks will find the method explained, with reference to the dodecahedron, in solutions to problems E1752 and E1897 in *The American Mathematical Monthly*, February 1966, page 200, and October 1967, pages 1008–10. For the rhombic dodecahedron, which is not regular, see O'Beirne's article "A Nonsense Result in the Traffic Statistics of Drunk Flies," in the *Bulletin of the Institute of Mathematics and Its Applications*, August 1966, pages 116–19.

Steps need not be to adjacent spots on a network. Consider the symmetric random ergodic walk of a rook over a chessboard, assuming that on each move the rook chooses between all possible moves with equal probability. Since a rook can reach any cell from 14 other cells, the transition probability for every move is 1/14. The rook will therefore spend the same time on each cell.

The situation with respect to other chess pieces is different because their transition probabilities vary. A knight, for example, can reach a corner cell only from two other cells, whereas it can reach any of the 16 central cells from eight other cells. Since the proportion here is 2/8 or 1/4, it follows that during an endless random walk over the chessboard, a knight will visit any designated corner square one-fourth as often as it will visit a given square among the central 16. For a proof see "Generalized Symmetric Random Walks," by Eugene Albert, in *Scripta Mathematica*, August 1964, pages 185–87.

ADDENDUM

EARLIER IN THIS CHAPTER we encountered a pleasant theorem that is expressed in graph-theoretic terms as follows. Consider any graph that is regular in the sense that every point belongs to the same number of lines. If a bug starts at any point and makes a random walk, at each point choosing one of the available lines with equal probability, the expected (average) number of steps back to the starting point is equal to the graph's number of points.

Although I gave some references that explain how to calculate such walks, I did not provide an example. It may be of interest to see how it is done for a triangle and tetrahedron. The reader may then be able to generalize the procedure to the edges of polygons and polyhedrons, and other regular graphs.

Label the corners of a triangle A, B, C. We wish to know the expected length of a random walk from A back to A, assuming that at each corner one of the two edges is chosen with equal probability. Note that this is the same as calling A an absorbing barrier (after the first step is made), then asking for the expected length of a random walk before the bug is absorbed.

Let x be the expected length of a walk from B to A. For symmetry reasons it is the same as the expected walk from C to A.

Suppose the bug is at B. If it chooses to walk to A, the expected path to A is 1. If it chooses to walk to C, the expected path to A is 1 plus the expected path from C to A. The latter is equal to x, therefore the expected path from C to A is $(1 + x)$. Add the lengths of the two paths, $1 + (1 + x)$, and divide by 2 to get the average. Thus we have the following simple equation:

$$x = \frac{1 + (1 + x)}{2},$$

which gives x a value of 2.

We now know that if the bug starts at either B or C, the expected path to A is 2. If the bug starts at A, however, it must go 1 step to get to either B or C. Therefore the expected length of a walk from A back to A is $1 + 2 = 3$.

The tetrahedron is easily solved the same way. Label the corners A, B, C, D. The bug starts at A. If it chooses to go directly to A, the expected path is 1. If it chooses to go to C or D, the expected path to A is $(1 + x)$. The average walk from B to A, therefore, is $1 + (1 + x) + (1 + x)$ divided by 3. Our equation is

$$x = \frac{1 + (1 + x) + (1 + x)}{3},$$

which gives x a value of 3. The bug at A must make 1 step to get to one of the other three corners, therefore the expected walk from A back to A is $1 + 3 = 4$.

As an exercise, with only slightly more complicated equations, the reader may enjoy proving that the expected walk on a square or cube, from a corner back to the same corner, is respectively 4 and 8.

ANSWERS

1. Two MEN start at the same spot on the plane. One makes a random walk of 70 unit steps, the other a random walk of 30 steps. What is the expected (average) distance between them at the finish? If you imagine one man reversing the direction of his walk until he returns to where he started and then continuing along the other man's path, you will see that the question is the same as asking for the expected distance from the starting spot of a single random walk of 100 steps. We were told that this expected distance is the average length of a step times the square root of the number of steps. Therefore the answer is 10 units.

2. Because of the cube's symmetry any first step of the drunken fly is sure to take it toward the cube's most distant corner, where the drunken spider has begun its simultaneous walk. It does not matter, therefore, what first step the fly takes. The spider, however, can reach a corner adjacent to the fly only by taking two of its three equally probable first steps. Therefore the probability that, after their first steps, the creatures will be at adjacent corners is 2/3. At every possible pair of adjacent corners they can then occupy, the probability that the fly will move toward the spider is 2/5 and the probability that the spider will move toward the fly is 1/4. The product of these three probabilities, 2/3, 2/5, and 1/4, is 1/15. This is the probability that the spider and the fly will meet in the middle of an edge after each has traveled 1½ edges.

CHAPTER 8

Boolean Algebra

ARISTOTLE deserves full credit as the founder of formal logic even though he restricted his attention almost entirely to the syllogism. Today, when the syllogism has become a trivial part of logic, it is hard to believe that for 2,000 years it was the principal topic of logical studies, and that as late as 1797 Immanuel Kant could write that logic was "a closed and completed body of doctrine."

"In syllogistic inference," Bertrand Russell once explained, "you are supposed to know already that all men are mortal and that Socrates is a man; hence you deduce what you never suspected before, that Socrates is mortal. This form of inference does actually occur, though very rarely." Russell goes on to say that the only instance he ever heard of was prompted by a comic issue of *Mind*, a British philosophical journal, that the editors concocted as a special Christmas number in 1901. A German philosopher, puzzled by the magazine's advertisements, eventually reasoned: Everything in this magazine is a joke, the advertisements are in this magazine, therefore the advertisements are jokes. "If you wish to become a logician," Russell wrote elsewhere, "there is one piece of sound advice which I cannot urge too strongly, and that is: Do *not* learn the traditional logic. In Aristotle's day it was a creditable effort, but so was the Ptolemaic astronomy."

The big turning point came in 1847 when George Boole (1815–1864), a modest, self-taught son of a poor English shoemaker [*see Figure 40*], published *The Mathematical Analysis of Logic*. This and other papers led to his appointment (although he had no university degree) as professor of mathematics at Queens College (now University College) at Cork in Ireland, where he wrote his treatise *An Investigation of the Laws of Thought, on Which are Founded the Mathematical Theories of Logic and Probabilities* (London, 1854). The basic idea— substituting symbols for all the words used in formal logic—had occurred to others before, but Boole was the first to produce a workable system. By and large, neither philosophers nor mathematicians of his century showed much interest in this remarkable achievement. Perhaps that was one reason for Boole's tolerant attitude toward mathematical eccentrics. He wrote an article about a Cork crank named John Walsh (*Philosophical Magazine*, November 1851) that Augustus De Morgan, in his *Budget of Paradoxes*, calls "the best biography of a single hero of the kind that I know."

Boole died of pneumonia at the age of 49. His illness was attributed to a chill that followed a lecture he gave in wet clothes after having been caught in the rain. He was survived by his wife and five daughters. Norman Gridgeman, writing "In Praise of Boole" (see bibliography), gives some fascinating details about the six ladies. Boole's wife, Mary Everest, wrote popular books about her husband's views on mathematics and education. One book is titled *The Philosophy and Fun of Algebra*. The oldest daughter, Mary, married Charles Hinton, a mathematician who wrote the first novel about flatland (see Chapter 12 of my *Unexpected Hanging*), as well as books on the fourth dimension.

FIGURE 40
George Boole

Margaret became the mother of Sir Geoffrey Taylor, a Cambridge mathematician. Alicia, intrigued by Charles Hinton's excursions into higher space dimensions, made some significant discoveries in the field. Lucy became a professor of chemistry. Ethel Lilian, the youngest daughter, married Wilfrid Voynich, a Polish scientist. They settled in Manhattan, where Ethel died in 1960. She wrote several novels, including one called *The Gadfly* (1898) that became so popular in Russia that three operas were based on it. In recent years a million copies have been sold in China. "Modern Russians are constantly amazed," writes Gridgeman, "that so few Westerners have heard of E. L. Voynich, the great English novelist."

The few who appreciated Boole's genius (notably the German mathematician Ernst Schröder) rapidly improved on Boole's notation, which was clumsy mainly because of Boole's attempt to make his system resemble traditional algebra. Today Boolean algebra refers to an "uninterpreted" abstract structure that can be axiomized in all kinds of ways but that is essentially a streamlined, simplified version of Boole's system. "Uninterpreted" means that no meanings whatever—in logic, mathematics, or the physical world—are assigned to the structure's symbols.

As in the case of all purely abstract algebras, many different interpretations can be given to Boolean symbols. Boole himself interpreted his system in the Aristotelian way as an algebra of classes and their properties, but he greatly extended the old class logic beyond the syllogism's narrow confines. Since Boole's notation has been discarded, modern Boolean algebra is now written in the symbols of set theory, a set being the same as what Boole meant by a class: any collection of individual "elements." A set can be finite, such as the numbers 1, 2, 3, the residents of Omaha who have green eyes, the corners of a cube, the planets of the solar system, or any other specified collection of things. A set also can be infinite, such as the set of even integers or possibly the set of all stars. If we specify a set, finite or infinite, and then consider all its subsets (they include the set itself as well as the empty set of no members) as being related to one another by inclusion (that is, the set 1, 2, 3 is included in the set 1, 2, 3, 4, 5), we can construct a Boolean set algebra.

A modern notation for such an algebra uses letters for sets, subsets, or elements. The "universal set," the largest set being considered, is symbolized by ∪. The empty, or null, set is ∅. The "union" of sets *a* and *b* (everything in *a* and *b*) is symbolized by ∪, sometimes called a cup. (The union of 1, 2 and 3, 4, 5 is 1, 2, 3, 4, 5.) The "intersection" of sets *a* and *b* (everything common to *a* and *b*) is symbolized by ∩, sometimes called a cap. (The intersection of 1, 2, 3 and 3, 4, 5 is 3.) If two sets are identical (for example, the set of odd numbers is the same as the set of all integers with a remainder of 1 when divided by 2), this is symbolized by =. The "complement" of set *a*—all elements of the universal set that are not in *a*—is indicated by *a'*. (The complement of 1, 2, with respect to the universal set 1, 2, 3, 4, 5, is 3, 4, 5.) Finally, the basic binary relation of set inclusion is symbolized by ε; *a* ε *b* means that *a* is a member of *b*.

As a matter of historical interest, Boole's symbols included letters for elements, classes, and subclasses: 1 for the universal class; 0 for the null class; + for class union (which he took in an "exclusive" sense to mean those elements of two classes that are *not* held in common; the switch to the "inclusive" sense, first made by the British logician and economist William Stanley Jevons, had so many advantages that later logicians adopted it); × for class intersection; = for identity; and the minus sign, −, for the removal of one set from another. To show the complement of *x*, Boole wrote 1 − *x*. He had no symbol for class inclusion but could express it in various ways such as *a* × *b* = *a*, meaning that the intersection of *a* and *b* is identical with all of *a*.

The Boolean algebra of sets can be elegantly diagrammed with Venn circles (after the English logician John Venn), which are now being introduced in many elementary school classes. Venn circles are diagrams of an interpretation of Boolean algebra in the point-set topology of the plane. Let two overlapping circles symbolize the union of two sets [*see Figure 41*], which we here take to be the set of the 10 digits and the set of the first 10 primes. The outer rectangle contains the universal set. This includes the area outside both circles, which is shaded to indicate that it is the null set; it is empty because we are con-

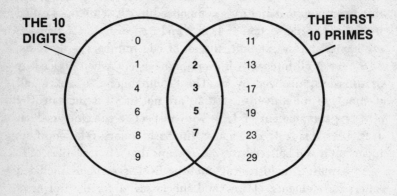

FIGURE 41
Venn diagram for set intersection

cerned solely with the elements inside the two circles. These 16 elements are the union of the two sets. The overlapping area contains the intersection. It consists of the set 2, 3, 5, 7: digits that are also among the first 10 primes.

Adopting the convention of shading any area known to represent an empty set, we can see how a three-circle Venn diagram proves the ancient syllogism Russell so scornfully cited. The circles are labeled to indicate sets of men, mortal things, and Socrates (a set with only one member). The first premise, "All men are mortal," is diagrammed by shading the men circle to show that the class of nonmortal men is empty [*see Figure 42, left*]. The second premise, "Socrates is a man," is similarly diagrammed by shading the Socrates circle to show that all of Socrates, namely himself, is inside the men circle [*see Figure 42, right*]. Now we inspect the diagram to see if the conclusion, "Socrates is mortal," is valid. It is. All of Socrates (the unshaded part of his circle marked by a dot) is inside the circle

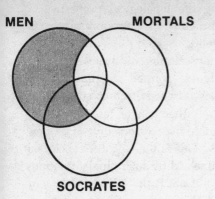

Premise: "All men are mortal."

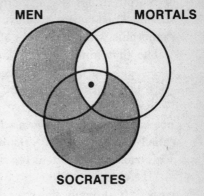

Premise: "Socrates is a man."

FIGURE 42

BOOLEAN SET ALGEBRA	PROPOSITIONAL CALCULUS
U (UNIVERSAL SET)	T (TRUE)
φ (NULL SET)	F (FALSE)
a, b, c, … (SETS, SUBSETS, ELEMENTS)	p, q, r, … (PROPOSITIONS)
a∪b (UNION: ALL OF a AND b)	p∨q (DISJUNCTION: EITHER p ALONE OR q ALONE, OR BOTH, ARE TRUE.)
a∩b (INTERSECTION: WHAT a AND b HAVE IN COMMON)	p•q (CONJUNCTION: BOTH p AND q ARE TRUE.)
a = b (IDENTITY: a AND b ARE THE SAME SET.)	p≡q (EQUIVALENCE: IF AND ONLY IF p IS TRUE, THEN q IS TRUE.)
a' (COMPLEMENT: ALL OF **U** THAT IS NOT a)	~p (NEGATION: p IS FALSE.)
a∈b (INCLUSION: a IS A MEMBER OF b.)	p⊃q (IMPLICATION: IF p IS TRUE, q IS TRUE.)

FIGURE 43

Corresponding symbols in two versions of Boolean algebra

of mortal things. By exploiting the topological properties of simple closed curves, we have a method of diagramming that is isomorphic with Boolean set algebra.

The first important new interpretation of Boolean algebra was suggested by Boole himself. He pointed out that if his 1 were taken as truth and his 0 as falsehood, the calculus could be applied to statements that are either true or false. Boole did not carry out this program but his successors did. It is now called the propositional calculus. This is the calculus concerned with true or false statements connected by such binary relations as "If p then q," "Either p or q but not both," "Either p or q or both," "If and only if p then q," "Not both p and q," and so on. The chart in Figure 43 shows the symbols of the propositional calculus that correspond to symbols for the Boolean set algebra.

It is easy to understand the isomorphism of the two interpretations by considering the syllogism about Socrates. Instead of saying, "All men are mortal," which puts it in terms of class properties or set inclusion, we rephrase it as, "If x is a man then x is a mortal." Now we are stating two propositions and joining them by the "connective" called "implication." This is diagrammed on Venn circles in exactly the same way we diagrammed "All men are mortal." Indeed, all the binary relations in the propositional calculus can be diagrammed with Venn circles and the circles can be used for solving simple problems in the calculus. It is shameful that writers of most introductory textbooks on formal logic have not yet caught on to this. They continue to use Venn circles to illustrate the old class-inclusion logic but fail to apply them to the propositional calculus, where they are just as efficient. Indeed, they are even more efficient, since in the propositional calculus one is unconcerned with the "existential quantifier," which asserts that a class is not empty because it has at least one member. This was expressed in the traditional logic by the word "some" (as in "Some apples are green"). To take care of such statements Boole had to tie his algebra into all sorts of complicated knots.

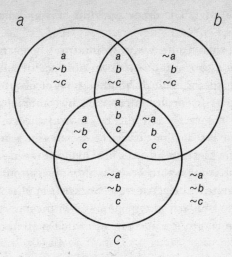

FIGURE 44
Venn diagram for martini puzzle

To see how easily the Venn circles solve certain types of logic puzzles, consider the following premises about three businessmen, Abner, Bill, and Charley, who lunch together every working day:

1. If Abner orders a martini, so does Bill.
2. Either Bill or Charley always orders a martini, but never both at the same lunch.
3. Either Abner or Charley or both always order a martini.
4. If Charley orders a martini, so does Abner.

To diagram these statements with Venn circles, we identify having a martini with truth and not having one with falsehood. The eight areas of the overlapping circles shown in Figure 44 are labeled to show all possible combinations of truth values for *a*, *b*, *c*, which stand for Abner, Bill, and Charley. Thus the area marked *a*, ~*b*, *c* represents Abner's and Charley's having martinis while Bill does not. See if you can shade the areas declared empty by the four premises and then examine the result to determine who will order martinis if you lunch with the three men.

There are many other ways to interpret Boolean algebra. It can be taken as a special case of an abstract structure called a ring, or as a special case of another type of abstract structure called a lattice. It can be interpreted in combinatorial theory, information theory, graph theory, matrix theory, and meta-mathematical theories of deductive systems in general. In recent years the most useful interpretation has been in switching theory, which is important in the design of electronic computers but is not limited to electrical networks. It applies to any kind of energy transmission along channels with connecting devices that turn the energy on and off, or switch it from one channel to another.

The energy can be a flowing gas or liquid, as in modern fluid control systems [see "Fluid Control Devices," by Stanley W. Angrist, in *Scientific American*, December 1964]. It can be light beams. It can be mechanical energy as in the logic machine Jevons invented for solving four-term problems in Boolean algebra. It can be rolling marbles, as in several computerlike toys now on the market: Dr. Nim, Think-a-Dot, and Digi-Comp II. And if inhabitants of another planet have a highly developed sense of smell, their computers could use odors transmitted through tubes to sniffing outlets. As long as the energy either moves or does not move along a channel, there is an isomorphism between the two states and the two truth values of the propositional calculus. For every binary connective in the calculus, there is a corresponding switching circuit. Three simple examples are shown in Figure 45. The bottom circuit is used whenever two widely separated electric light switches are used to control one light. It is easy to see that if the light is off, changing the state of either switch will turn it on, and if the light is on, either switch will turn it off.

This electrical-circuit interpretation of Boolean algebra had been suggested in a Russian journal by Paul S. Ehrenfest as

"AND" CIRCUIT: BULB LIGHTS ONLY
IF BOTH *a* AND *b* ARE CLOSED.

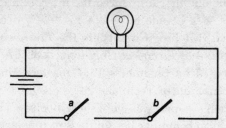

INCLUSIVE "OR" CIRCUIT: BULB LIGHTS ONLY
IF *a* OR *b* OR BOTH ARE CLOSED.

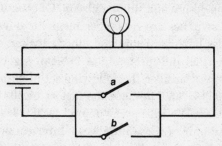

EXCLUSIVE "OR" CIRCUIT: BULB LIGHTS ONLY
IF *a* OR *b*, BUT *NOT* BOTH, IS LOWERED.

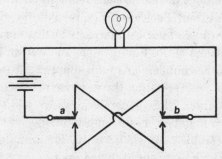

FIGURE 45
Circuits for three binary relations

early as 1910 and independently in Japan in 1936, but the first major paper, the one that introduced the interpretation to computer designers, was Claude E. Shannon's "A Symbolic Analysis of Relay and Switching Circuits" in the *Transactions of the American Institute of Electrical Engineers*, Vol. 57, December 1938. It was based on Shannon's 1937 master's thesis at the Massachusetts Institute of Technology.

Since Shannon's paper was published, Boolean algebra has become essential to computer design. It is particularly valuable in simplifying circuits to save hardware. A circuit is first translated into a statement in symbolic logic, the statement is "minimized" by clever methods, and the simpler statement is translated back to the design of a simpler circuit. Of course, in modern computers the switches are no longer magnetic devices or vacuum-tube diodes but transistors and other tiny semiconductors.

Now for one final interpretation of Boolean algebra that is a genuine curiosity. Consider the following set of eight numbers: 1, 2, 3, 5, 6, 10, 15, 30. They are the factors of 30, including 1 and 30 as factors. We interpret "union" as the least common multiple of any pair of those numbers. "Intersection" of a pair is taken to be their greatest common divisor. Set inclusion becomes the relation "is a factor of." The universal set is 30, the null set 1. The complement of a number a is $30/a$. With these novel interpretations of the Boolean relations, it turns out that we have a consistent Boolean structure! All the theorems of Boolean algebra have their counterparts in this curious system based on the factors of 30. For example, in Boolean algebra the complement of the complement of a is simply a, or in the propositional-calculus interpretation the negation of a negation is the same as no negation. More generally, only an odd series of negations equals a negation. Let us apply this Boolean law to the number 3. Its complement is $30/3 = 10$. The complement of 10 is $30/10 = 3$, which brings us back to 3 again.

Consider two famous Boolean laws called De Morgan's laws. In the algebra of sets they are

$$(a \cup b)' = a' \cap b'$$
$$(a \cap b)' = a' \cup b'.$$

In the propositional calculus they look like this:

$$\sim (a \vee b) \equiv \sim a \cdot \sim b$$
$$\sim (a \cdot b) \equiv \sim a \vee \sim b.$$

If the reader will substitute any two factors of 30 for a and b, and interpret the symbols as explained, he will find that De Morgan's laws hold. The fact that De Morgan's laws form a pair illustrates the famous duality principle of Boolean algebra. If in any statement you interchange union and intersection (if and wherever they appear) and interchange the universal and the null sets, and also reverse the direction of set inclusion, the result is another valid law. Moreover, these changes can be made all along the steps of the proof of one law to provide a valid proof of the other! (An equally beautiful duality principle holds in projective geometry with respect to interchanges of lines and points.)

The numbers 1, 2, 3, 5, 6, 7, 10, 14, 15, 21, 30, 35, 42, 70, 105, 210—the 16 factors of 210—also form a Boolean algebra when interpreted in the same way, although of course 210 is now the universal set and the complement of a is $210/a$. Can the reader discover a simple way to generate sets of 2^n numbers, where n is any positive integer, that will form Boolean systems of this peculiar kind?

ANSWERS

THREE VENN CIRCLES are shaded as in Figure 46 to solve the problem about the three men who lunch together. Each of the first four diagrams is shaded to represent one of the four premises of the problem. Superimposing the four to form the last diagram shows that if the four premises are true, the only possible

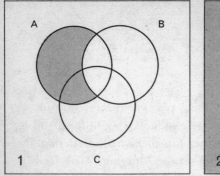

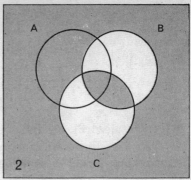

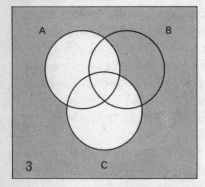

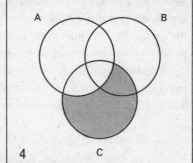

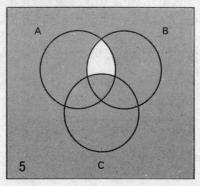

FIGURE 46

Venn-diagram solution to martini problem

combination of truth values is a, b, $\sim c$, or true a, true b, and false c. Since we are identifying truth with ordering a martini, this means that Abner and Bill always order martinis, whereas Charley never does.

The method of generating 2^n integers to form Boolean algebras was given by Francis D. Parker in *The American Mathematical Monthly* for March 1960, page 268. Consider a set of any number of distinct primes, say 2, 3, 5. Write down the multiples of all the subsets of these three primes, which include 0 (the null set) and the original set of three primes. Change 0 to 1. This produces the set 1, 2, 3, 5, 6, 10, 15, 30, the first of the examples given. In a similar way the four primes 2, 3, 5, 7 will generate the second example, the $2^4 = 16$ factors of 210. A proof that all such sets provide Boolean algebras can be found in *Boolean Algebra*, by R. L. Goodstein in the answer to problem No. 10, page 126.

CHAPTER 9

Can Machines Think?

There was a time when it must have seemed highly improbable that machines should learn to make their wants known by sound, even through the ears of man; may we not conceive, then, that a day will come when those ears will be no longer needed, and the hearing will be done by the delicacy of the machine's own construction?—when its language shall have been developed from the cry of animals to a speech as intricate as our own?

—SAMUEL BUTLER, *Erewhon*

ALAN MATHISON TURING, a British mathematician who died in 1954 at the age of 42, was one of the most creative of the early computer scientists. Today he is best known for his concept of the Turing machine. We shall take a quick look at such machines and then consider one of Turing's less well-known ideas, the Turing game— a game that leads to deep and unsettled philosophical controversies.

A Turing machine is a "black box" (a machine with unspecified mechanisms) capable of scanning an infinite tape of square cells. The box can have any finite number of states. A finite portion of the tape consists of nonblank cells, each bearing one of a finite number of symbols. When the box views a cell, it can leave a symbol unaltered, erase it, erase it and print another symbol, or print a symbol in a blank cell. The tape is then shifted one cell to the left or right or stays fixed; the box either remains in the same state or clicks to a different state.

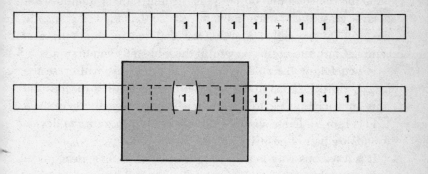

	STATE A	STATE B
1	1. ERASE THE 1. 2. SCAN NEXT CELL ON RIGHT. 3. GO TO STATE B.	1. SCAN NEXT CELL ON RIGHT. 2. STAY IN STATE B.
+		1. ERASE THE +. 2. PRINT 1. 3. STOP.

FIGURE 47
A Turing machine for addition

A table of rules describes what the box does for every achievable combination of symbol and state. Such a table completely defines a particular Turing machine. There is a countable (aleph null) infinity of Turing machines, each designed for a specific task, and for every task the machine's structure can vary widely in symbols, states, and rules.

A good way to grasp the essence of a Turing machine is to make one, albeit a trivial one [*see Figure 47*]. Eight cells on the paper tape are marked $1111 + 111$, signifying the addition of 4 and 3 in the "unary" system in which an integer n is symbolized by n 1's. To make the machine, draw a small square (the black box) and cut two slits in it so that the tape can be inserted as shown. Adjust the tape so that the first 1 is visible. The table at the bottom of the picture gives all the necessary rules.

Start by assuming that the machine is in state A. Consult the table for the combination of symbol 1 and state A and do what it says: erase the 1, move the tape left (so that the box scans the next cell to the right) and assume that the machine clicks to state B. Continue in this way until the table tells you to stop.

If you follow the rules correctly, the machine will erase the first 1, shift the tape to the left cell by cell until it reaches the plus sign, change $+$ to 1 and stop. The strip will then show 1111111, or 7. These simple rules obviously program the device to add any pair of positive integers, however large.

It is a tedious way to add, of course, but Turing's idea was to reduce machine computation to a simple and abstract schema, making it easier to analyze all kinds of thorny theoretical problems, such as what can and what cannot be computed. Turing showed that his idealized device can be programmed to do, in its clumsy way, anything the most powerful electronic computer can do. Like any computer—and like the human brain—it is limited by the fact that certain calculations (such as calculating pi) require an infinite number of steps and by the fact that some problems are unsolvable in principle; there *is* no algorithm, or effective procedure, by which they can be solved. A "universal Turing machine" is capable of doing whatever any special-purpose Turing machine can do. In brief, it computes anything that is computable.

In 1950 Turing's article "Computing Machinery and Intelligence" appeared in *Mind*, a British philosophical journal, and it has since been reprinted in several anthologies, including James R. Newman's *The World of Mathematics*. "I propose," Turing began, "to consider the question, 'Can machines think?'" This, Turing decided, was much too vague to have a meaningful answer. He proposed instead a related but more precise question: Can a computer be taught to win the "imitation game," now commonly called the Turing game or Turing test?

Turing based his test on a parlor game in which a man is concealed in one room and a woman in another. An interrogator of either sex asks the concealed players questions, which are conveyed by an intermediary; the answers are returned in typescript. Each player tries to convince the interrogator that he or she is, say, the woman. The interrogator wins if he guesses correctly who is telling the truth.

Suppose, Turing said, we replace one player with a learning machine that has been taught to converse in an ordinary language such as English. Is it possible for such a machine to deceive an interrogator when both the machine and its human partner try to persuade the questioner that he, she, or it is the human?

Several continuums blur the meaning of "deceive." How long a conversation is allowed? How intelligent is the interrogator? How intelligent is the person competing against the machine? It is possible today for a computer to pass the Turing test if the questioner is a child and is allowed only a few questions. There may be no dramatic future breakthroughs, just as there probably were no dramatic breakthroughs in the evolution of man. Conversation machines could improve gradually, so that it takes increasingly long dialogues, with increasingly clever interrogators, to beat the machines. Perhaps someday only an interrogating computer will consistently guess correctly. Turing himself made a cautious prediction. By the year 2000, he wrote, computers will speak fluently enough to deceive an "average interrogator" at least 30 percent of the time after "about five minutes" of dialogue. He imagined the following conversation as being typical:

Q. Please write me a sonnet on the subject of the Forth Bridge.

A. Count me out on this one. I never could write poetry.

Q. Add 34957 to 70764.

A. (Pause about 30 seconds and then give as answer) 105621.

Q. Do you play chess?

A. Yes.

Q. I have K at my K1, and no other pieces. You have only K at K6 and R at R1. It is your move. What do you play?

A. (After a pause of 15 seconds) R–R8 mate.

Many programs enabling computers to simulate English conversation now exist. Joseph Weizenbaum, an electrical engineer at the Massachusetts Institute of Technology, has written one he calls Eliza, after Eliza Doolittle of Shaw's *Pygmalion*. *The New York Times* (June 15, 1968) reported a typewritten dialogue with a woman in which an IBM 7094 computer assumed the role of a psychiatrist:

W. Men are all alike.

C. In what way?

W. They're always bugging us about something or other.

C. Can you think of a specific example?

W. Well, my boyfriend made me come here.

C. Your boyfriend made you come here?

W. He says I'm depressed much of the time.

C. I am sorry to hear you are depressed.

W. It's true. I am unhappy.

C. Do you think coming here will help you not to be unhappy?

W. I need some help, that much seems certain.

C. What would it mean to you if you got some help?

W. Perhaps I could learn to get along with my mother.

C. Tell me more about your family.

W. My mother takes care of me.

C. Who else in your family takes care of you?

W. My father.

C. Your father?

W. You are like my father in some ways.

C. What resemblance do you see?

W. You are not very aggressive but I think you don't want me to notice that.

c. What makes you think I am not very aggressive?
w. You don't argue with me.

The dialogue continues in a similar vein and is indistinguishable from a conversation between a patient and a nondirective therapist. The program was not even a learning program. Weizenbaum readily admitted that the computer did not "understand" anything being said. It certainly could not pass the Turing test. Suppose, however, that by the year 2000 computers can play the Turing game as well as they now play checkers and chess. What, if anything, will this reveal about the nature of the machine's "mind"?

Readers of Arthur C. Clarke's novel *2001: A Space Odyssey* may recall that HAL, the spaceship's talking computer, is said to "think" because he could "pass the Turing test with ease." (HAL stands for *h*euristically programmed *al*gorithmic computer, but Clarke may have had some trickier wordplay in mind when he picked the name. Can the reader figure out what it is?) Does HAL really think or does he just mimic thinking? Turing believed that when the time comes that computers converse well enough to pass his test, no one will hesitate to say that they are thinking.

Enormously tangled questions immediately arise. Can such a computer be self-conscious? Can it have emotions? A sense of humor? In short, should it be called a "person" or just a dead machine built to imitate a person? L. Frank Baum described Tiktok, a windup mechanical man, as a robot that "thinks, speaks, acts, and does everything but live."

Surely the ability of a computer to pass Turing tests would prove only that a computer could imitate human speech well enough to pass such tests. Suppose someone in the Middle Ages had proposed the following "tulip test." Will it ever be possible to make an artificial tulip that cannot be distinguished from a real tulip if one is allowed only to look at it? Fake tulips can now pass this test, but this tells us nothing about a chemist's ability to synthesize organic compounds or to make a tulip that will grow like a tulip. Just as we can touch what we think is a flower and exclaim, "Oh—it's artificial!" it seems unsurprising that a day might come when we can hold a lengthy conversation with what we think is a person, then open a door and be amazed to discover that we have been talking to a computer.

Keith Gunderson, in an important 1964 paper in which he criticized Turing for making too much of the significance of his test, expressed the point this way. "In the end, the steam drill outlasted John Henry as a digger of railway tunnels, but that didn't prove the machine had muscles; it proved that muscles were not necessary for digging railway tunnels."

A curious twist was given to the Turing test in a lecture by Michael Scriven, reprinted as "The Compleat Robot: A Prolegomena to Androidology" in *Dimensions of Mind*, edited by Sidney Hook. Scriven conceded that conversational ability does not prove that a computer possesses other attributes of a "person." Suppose, however, a conversing computer is taught the meaning of "truth" (in, say, the correspondence sense made precise by Alfred Tarski) and then is programmed so that it cannot lie. "This makes the robot unsuitable," Scriven said, "for use as a personal servant, advertising copywriter, or politician, but renders it capable of another service." We can now ask if it is aware that it exists, has emotions, thinks some jokes are funny, acts on its own free will, enjoys Keats and so on, and expect it to give correct answers.

There is the possibility that a "Scriven machine" (as it is called by one of several philosophers who criticize Scriven's paper in other chapters of Hook's anthology) will say no to all such questions. But if it gives yes answers, Scriven contends, we have as much justification for believing it as we have for believing a human, and no reason for not calling it a "person."

Philosophers disagree about Turing's and Scriven's arguments. In a short piece titled "The Supercomputer as Liar," Scriven replied to some of his critics. Mortimer J. Adler, in his book *The Difference of Man and the Difference It Makes*, takes the view that the Turing test is an "all-or-none affair," and that success or continued failure in creating computers capable of passing it will respectively weaken or strengthen the view that a man is radically different in kind from any possible machine as well as any subhuman animal.

Would conversing machines really alter the beliefs of people who hold such a view? It is not hard to imagine a television show 50 years from now in which guests ad-lib with a robot Johnny Carson whose memory has been stocked with a million jokes and that has been taught the art of timing by human comics. I doubt that anyone would suppose the computer had a sense of humor any more than a person defeated by a robot chess player supposes he has played against a machine radically different in kind from a computer that plays ticktacktoe. Rules of syntax and semantics are just not all that different from rules of chess.

At any rate, the debate continues, complicated by metaphysical and religious commitments and complex linguistic problems. All the age-old enigmas about mind and body and the nature of personality are being reformulated in a new terminology. It is hard to predict what thresholds will be crossed and how the crossings will affect fundamental philosophical disagreements as robots of the future improve—as they surely will—in their ability to think, speak, and act like humans.

Samuel Butler's chapters in *Erewhon* explaining why the Erewhonians destroyed their machines before the machines could become masters instead of servants were read 100 years ago as far-fetched satire. Today they read like sober prophecy. "There is no security," Butler wrote, "against the ultimate development of mechanical consciousness, in the fact of machines possessing little consciousness now. A mollusc has not much consciousness. Reflect upon the extraordinary advance which machines have made during the last few hundred years, and note how slowly the animal and vegetable kingdoms are advancing. The more highly organized machines are creatures not so much of yesterday, as of the last five minutes, so to speak, in comparison with past time."

ANSWERS

IF EACH LETTER in HAL is shifted forward one letter in the alphabet, the result is IBM. Because the IBM logo is visible on HAL's display terminals, everybody assumed that the letter shift was intentional on Arthur Clarke's part. Clarke has since assured me that it was totally accidental, and that he was astounded when the shift was first called to his attention.

CHAPTER 10

Cyclic Numbers

THE NUMBER 142,857, which students of recreational number theory are likely to recognize at once, is one of the most remarkable of integers. Apart from the trivial case of 1, it is the smallest "cyclic number." A cyclic number is an integer of n digits with an unusual property: when multiplied by any number from 1 through n, the product contains the same n digits of the original number in the same cyclic order. Think of 142,857 as being joined end to end in a circular chain. The circle can be broken at six points and the chain can be opened to form six six-digit numbers, the six cyclic permutations of the original digits:

$$1 \times 142{,}857 = 142{,}857$$
$$2 \times 142{,}857 = 285{,}714$$
$$3 \times 142{,}857 = 428{,}571$$
$$4 \times 142{,}857 = 571{,}428$$
$$5 \times 142{,}857 = 714{,}285$$
$$6 \times 142{,}857 = 857{,}142$$

The cyclic nature of the six products has long intrigued magicians, and many clever mathematical prediction tricks are based on it. Here is one:

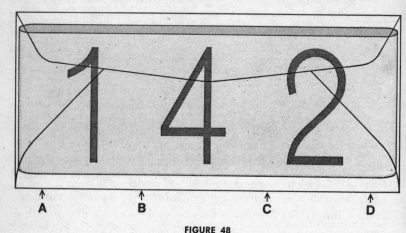

FIGURE 48
*Endless strip (top) placed in envelope (bottom)
for prediction trick*

Prepare a deck of playing cards by removing the nine spades with digit values. Place them on the bottom of the deck so that their order from the bottom up is 1, 4, 2, 8, 5, 7, with the remaining three cards following in any order. Your prediction of the trick's outcome is the number 142,857 written in large numerals on a strip of paper twice as long as the envelope into which it will be put. Paste the ends of the strip together to make a circular band, with the numerals on the outside, then press the band flat as shown in Figure 48. Thus flattened, the band is sealed into the envelope as shown.

You have, of course, memorized the number 142,857, recalling also that its first three digits are on the top side of the band, its last three digits on the bottom side. The envelope later is cut open by scissoring it at one of the four places marked *A, B, C, D*. If you cut at *A* or *D*, cut through the end of the band also, so that when it is pulled from the envelope, it will be a strip bearing 142,857 or 857,142. The other four permutations are obtained by cutting the envelope at *B* or *C*. Start the cut in the envelope only, below the band. As you continue cutting, make sure that the scissors snip only the upper part of the band and the upper part of the envelope. In this way you can pull from the slit a strip bearing 428,571 or 285,714. For the remaining two permutations simply turn the envelope over and follow the same procedure on the other side. This way of cutting an envelope to take from it a strip showing any of the six cyclic permutations of 142,857 is based on a procedure devised by Samuel Schwartz, a New York City attorney and amateur magician. He uses a window envelope so that spectators can see numerals on the band, and a slightly different preparation and handling, but the method is essentially the same.

The sealed envelope with your prediction is given to someone at the start of the trick. Hand another person the prepared deck and ask him to give it two thorough riffle shuffles (the usual kind of shuffle in which the deck is separated into two piles and the piles are interleaved). The two shuffles will distribute the nine spades upward through the deck without disturbing their order. To obtain a random six-digit number, you explain, you will go through the shuffled deck, cards face up, and remove the first six spades that have digit values. The digits will be 1, 4, 2, 8, 5, 7. Arrange the six cards in a row on a table. A random multiplier from 1 through 6 is now obtained by rolling a die. Better still, hand someone an imaginary die and ask him to roll this invisible die and tell you what he "sees" on top. Multiply 142,857 by the digit he names. Cut open the envelope properly (in order to determine where to cut, multiply 7 by the selected digit to get the last digit of the product) and pull out the strip to prove you have correctly predicted the product.

The number 142,857 is involved in many other mathematical magic tricks (see bibliography for several references). A defect of all these tricks is that spectators may notice that the same digits of 142,857 reappear in the prediction; also, the magic number itself has now become fairly well known. One way around this is to use, instead of 142,857, the quotient obtained by dividing it by one of its factors. For example: $142,857/3 = 47,619$. Instead of having 47,619 multiplied by 1, 2, 3, 4, 5, or 6, have it multiplied by any of the first six multiples of 3. The result, of course, will be a cyclic permutation of 142,857. Or you can use $142,857/9 = 15,873$ and have it multiplied by any of the first six multiples of 9; or $142,857/11 = 12,987$ and have it multiplied by 11, 22, 33, 44, 55, or 66; and so on.

Many centuries ago, when mathematicians first became aware of the cyclic character of 142,857, they began looking for larger numbers with the same whimsical property. Early work along such lines is summarized in the first volume of Leonard Eugene Dickson's *History of the Theory of Numbers*, Chapter 6, and dozens of papers on the topic have been written since Dickson's history first appeared in 1919. It turns out that all cyclic numbers are the periods (sometimes called repetends) in the repeating decimals (also called recurring, circulating, or periodic decimals) of the reciprocals of certain prime numbers. The reciprocal of 7, or 1/7, produces the repeating decimal .142,857; 142,857; 142,857. . . . Note that the number of digits in the period is one less than 7, which generates it. This provides one way of finding higher cyclics. If $1/p$, where p is a prime, produces a repeating decimal with a period of $p - 1$ digits, the period is a cyclic number. The next-larger prime that generates such a number is 17. Its repeating period is the 16-digit cyclic 0,588,235,294,117,647. Multiplied by any number from 1 through 16, the product repeats those 16 digits in the same cyclic order. All cyclics generated by primes higher than 7 must

begin with one or more 0's. If these numbers are used for prediction tricks or lightning-calculation stunts, the initial 0's can
be dropped—provided, of course, that you remember to insert
them at the proper place in the final product.

Exactly nine primes smaller than 100 generate cyclic numbers: 7, 17, 19, 23, 29, 47, 59, 61, 97. In the 19th century many
larger cyclics were found. William Shanks, who is best known
for his flawed calculation of pi to 707 decimals, discovered the
cyclic number generated by 1/17,389 and calculated its 17,388
digits (correctly).

No fraction with a denominator d can have a repeating period longer than $d - 1$ digits. Since this maximum-length period is achieved only when d is a prime, it follows that cyclic
numbers are equivalent to periods of maximum length for reciprocals of an integer. It is easy to see why $d - 1$ gives the longest possible period. When 1.000 . . . is divided by d, there are
only $d - 1$ possible remainders at each step of the division
process. As soon as a remainder is repeated, the period will start
over, and therefore no fraction with a denominator d can have
a period longer than $d - 1$ digits. It is also easy to see why such
maximum-length periods are cyclic. Consider 8/17. Since every
possible remainder appears in dividing 1 by 17, dividing 8 by
17 is merely starting the cyclic process at a different place. You
are certain to get the same cyclic order of digits in the period of
the repeating decimal. Multiplying the cyclic number generated by 1/17 by 8 is the same as finding the period for 8/17;
consequently the product must be a cyclic permutation of the
same 16 digits in the period for 1/17.

No nonrecursive formula is known that will automatically
generate all primes with reciprocals of maximum period length
(and hence generate all cyclic numbers), but there are many
dodges that simplify the identification of such primes and the
computer search programs for them. It is not yet known if
there are infinitely many primes that generate cyclic numbers,
but the conjecture seems highly probable. In Samuel Yates'

valuable table of prime period lengths for all primes through 1,370,471 (see bibliography), about 3/8 of the primes are of this type. The proportion remains fairly constant throughout, and there is a reasonable conjecture that it holds for all primes.

When a cyclic number is multiplied by its generating prime, the product is always a row of 9's. For instance, 142,857 times 7 is 999,999. This provides another way to search for cyclics: divide a prime, p, into a row of 9's until there is no remainder. If the quotient has $p - 1$ digits, it is a cyclic number. Even less expected is the fact that every cyclic (or any of its cyclic permutations), when split in half, gives two numbers that add to a row of 9's. For example, $142 + 857 = 999$. For another example, split the cyclic generated by $1/17$ into halves and add:

$$
\begin{array}{r}
05,882,352 \\
94,117,647 \\
\hline
99,999,999
\end{array}
$$

This surprising property is a special case of "Midy's theorem," credited by Dickson to E. Midy, who published it in France in 1836. The theorem states that if the period of a repeating decimal for a/p (where p is a prime and a/p is reduced to its lowest terms) has an even number of digits, the sum of its two halves will be a string of 9's. Some primes, such as 11, have periods of even length that are not cyclic numbers and yet have the 9's property. Other primes, such as 3 and 31, have periods of odd length. All cyclic numbers are even in length, however, and therefore Midy's theorem applies to them. This is good to remember because if you are doing the division to obtain a cyclic, you need only proceed halfway. You can write the remaining digits quickly by considering the digits already obtained and simply putting down their differences from nine. Of course it follows at once from Midy's theorem that all cyclics are multiples of nine, because any number whose digits add to a multiple of nine must itself be a multiple of nine. Readers interested in an elementary proof of Midy's theorem can find it in *The*

Enjoyment of Mathematics, by Hans Rademacher and Otto Toeplitz (Princeton University Press, 1957), pages 158–60. Another proof is given by W. G. Leavitt in "A Theorem on Repeating Decimals" in *The American Mathematical Monthly*, June–July 1967, pages 669–73.

There are many other strange properties of cyclic numbers, some of which the reader may have the pleasure of discovering for himself. I shall mention only one more. Every cyclic can be generated in numerous ways as the sum of an infinite arithmetic progression written diagonally. For example, start with 14 and double at each step, writing every number so that it projects two digits to the right:

$$
\begin{array}{l}
14 \\
\quad 28 \\
\qquad 56 \\
\qquad\quad 112 \\
\qquad\qquad 224 \\
\qquad\qquad\quad 448 \\
\qquad\qquad\qquad 896 \\
\qquad\qquad\qquad\qquad \cdot \\
\qquad\qquad\qquad\qquad\quad \cdot \\
\qquad\qquad\qquad\qquad\qquad \cdot \\
\hline
\end{array}
$$

142857142857 . . .

The sum repeats the smallest nontrivial cyclic number. Another way to obtain the same cyclic is to start with 7 and move diagonally down and left, multiplying by five at each step and keeping the diagonal uniform on the right:

$$
\begin{array}{r}
7 \\
35 \\
175 \\
875 \\
4375 \\
\cdot \\
\cdot \\
\cdot \\
\hline
\end{array}
$$

. . . 142857

This same procedure, using the simple doubling series 1, 2, 4, 8, 16, 32, . . . , will give the period of 1/19, the third cyclic number: 052,631,578,947,368,421. The tripling series 01, 03, 09, 27, 81, . . . , written diagonally down and right, each step projecting two digits rightward, has a sum that repeats the period of 1/97, the largest cyclic generated by a prime smaller than 100.

I conclude this brief discussion, which covers only a small portion of the fascinating properties of cyclics, by asking the reader what cyclic properties he can discover for the period of 1/13. This period, 076,923, is not a true cyclic. It can be called a cyclic of order-2. The answer will open a new field for exploration that is closely tied up with the better-known order-1 cyclics we have been considering.

ADDENDUM

JOHN W. WARD called my attention to the perfect magic square shown in Figure 49. It appears on page 176 of W. S. Andrews, *Magic Squares and Cubes,* a 1917 work currently available as a Dover paperback. The square is based on the cyclic number obtained from 1/19. All rows, columns, and main diagonals add to 81.

It is readily apparent that any cyclic number will provide a square that is magic for rows and columns, but Ward found that the one shown here is unique in also including the two main diagonals. As Andrews put it, "It is not easy to understand why each of the two diagonals of this square should sum 81, but if they are written one over the other, each pair of numbers will sum 9." Ward also showed that all semiperfect magic squares based on cyclic numbers have diagonals whose sums add to twice the magic constant.

Many readers wondered about and investigated what happens when a cyclic number of the first order is multiplied by

$1/19$ =	.0	5	2	6	3	1	5	7	8	9	4	7	3	6	8	4	2	1
$2/19$ =	.1	0	5	2	6	3	1	5	7	8	9	4	7	3	6	8	4	2
$3/19$ =	.1	5	7	8	9	4	7	3	6	8	4	2	1	0	5	2	6	3
$4/19$ =	.2	1	0	5	2	6	3	1	5	7	8	9	4	7	3	6	8	4
$5/19$ =	.2	6	3	1	5	7	8	9	4	7	3	6	8	4	2	1	0	5
$6/19$ =	.3	1	5	7	8	9	4	7	3	6	8	4	2	1	0	5	2	6
$7/19$ =	.3	6	8	4	2	1	0	5	2	6	3	1	5	7	8	9	4	7
$8/19$ =	.4	2	1	0	5	2	6	3	1	5	7	8	9	4	7	3	6	8
$9/19$ =	.4	7	3	6	8	4	2	1	0	5	2	6	3	1	5	7	8	9
$10/19$ =	.5	2	6	3	1	5	7	8	9	4	7	3	6	8	4	2	1	0
$11/19$ =	.5	7	8	9	4	7	3	6	8	4	2	1	0	5	2	6	3	1
$12/19$ =	.6	3	1	5	7	8	9	4	7	3	6	8	4	2	1	0	5	2
$13/19$ =	.6	8	4	2	1	0	5	2	6	3	1	5	7	8	9	4	7	3
$14/19$ =	.7	3	6	8	4	2	1	0	5	2	6	3	1	5	7	8	9	4
$15/19$ =	.7	8	9	4	7	3	6	8	4	2	1	0	5	2	6	3	1	5
$16/19$ =	.8	4	2	1	0	5	2	6	3	1	5	7	8	9	4	7	3	6
$17/19$ =	.8	9	4	7	3	6	8	4	2	1	0	5	2	6	3	1	5	7
$18/19$ =	.9	4	7	3	6	8	4	2	1	0	5	2	6	3	1	5	7	8

FIGURE 49

The only perfect magic square generated by a cyclic number

numbers larger than its number of digits, n. It turns out, pleasantly enough, that in all such cases the product can be reduced either to a cyclic permutation of the original number or to a number consisting of n nines. I will illustrate this with 142,857, and it will be apparent how it generalizes to larger cyclics.

We consider first all multipliers higher than n that are not multiples of $n + 1$. For example: $142,857 \times 123 = 17,571,411$. Mark off six digits from the right, and to this number add the number that remains:

$$
\begin{array}{r}
571411 \\
17 \\
\hline
571428
\end{array}
$$

The sum is a cyclic permutation of 142857.

Another example: $142,857^2 = 20,408,122,449$.

$$
\begin{array}{r}
20408 \\
122449 \\
\hline
142857
\end{array}
$$

If the multiplier is very large, we start from the right and partition the number into groups of six digits each. For example: $142,857 \times 6,430,514,712,336$.

$$
\begin{array}{r}
712336 \\
430514 \\
6 \\
\hline
1142856
\end{array}
$$

Because the sum has more than six digits, we must repeat the procedure:

$$
\begin{array}{r}
142856 \\
1 \\
\hline
142857
\end{array}
$$

If the multiplier of the cyclic number is a multiple of $n + 1$ (where n is the number of digits in the cyclic number), the procedure just described will produce a row of nines. For example: $142,857 \times 84 = 11,999,988$.

$$
\begin{array}{r}
999988 \\
11 \\
\hline
999999
\end{array}
$$

The reader can easily discover for himself how the procedure also applies to higher-order cyclic numbers.

Ratner's Star, a novel by Don DeLillo (Knopf, 1976), is spotted with references to 142,857 and its many strange numerological properties. The protagonist is a 14-year-old mathematical prodigy from the Bronx named Billy Twillig. He is hired by the government in 1979 for a top secret project that is trying to determine why a distant star in the Milky Way galaxy is sending 14–28–57 to the earth as a pulsed code. It finally turns out that the number means—but you'd best read the novel to find out.

ANSWERS

THE PERIOD OF $1/13$—076,923—is not a true cyclic in the sense defined previously. It is cyclic, however, in the following double sense. If multiplied by any number from 1 through 12, half of the products are the six cyclic permutations of 076,923 and the other half are the six cyclic permutations of 153,846. Note that each of these two numbers (like the smallest order-1 cyclic, 142,857) can be split in half and added to get 999. Moreover, each can be trisected and added to get 99 ($07 + 69 + 23 = 99$; $15 + 38 + 46 = 99$; $14 + 28 + 57 = 99$).

When a number with n digits, multiplied by 1 through $2n$, yields products that are all the cyclic permutations of two n-digit numbers, it is called an order-2 cyclic. Disregarding the trivial case generated by $1/3$, the lowest prime generating an order-2 cyclic is 13. Other primes under 100 that generate order-2 cyclics are 31, 43, 67, 71, 83 and 89.

Cyclic numbers can be of any order. The smallest prime generating an order-3 cyclic is 103. The repeating period of $1/103$, multiplied by any integer from 1 through 102, gives products that fall into three sets, each containing 34 cyclic permutations of a 34-digit number. The smallest prime generating an order-4 cyclic is 53. In general, as H. J. A. Dartnall, a London correspondent, has pointed out, if the reciprocal of a prime p has a repeating decimal period with a length (number of digits) equal to $(p - 1)/n$, the period is an n-order cyclic. For example, $1/37$ generates the three-digit period 027. Since $36/3 = 12$, the period is a 12-order cyclic. The lowest primes generating cyclics of orders 5 through 15 are respectively 11, 79, 211, 41, 73, 281, 353, 37, 2393, 449, 3061. Note that the 10 products of the order-5 cyclic, 09 (the period of 1/11), are the first 10 multiples of 9.

There is a curious relationship between the Fibonacci numbers and the order-2 cyclic generated by 89. Starting with 0, write the Fibonacci sequence in the diagonal form shown below, keeping the diagonal uniform on the right:

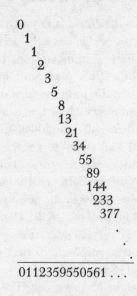

0112359550561 . . .

The sum is the reciprocal of 89.

The topic of higher-order cyclic numbers is a vast jungle, with a scattered literature that deserves to be brought together someday in a comprehensive bibliography. The same remarks apply to cyclic numbers in base systems other than 10. Every base notation has such numbers. In the binary, for example, the sequence of first-order cyclics (written in decimal notation) begins: 3, 5, 11, 13, 19, 29. . . .

CHAPTER 11

Eccentric Chess and Other Problems

1. ECCENTRIC CHESS

ON A RECENT VISIT to an imaginary chess club I found a game in progress between Mr. Black and Mr. White, the club's two most eccentric players. To my surprise the board appeared as shown in Figure 50. My first thought was that each player had started without his king's knight and that Black had moved first, but Mr. Black informed me that he had just completed his fourth move in a standard game that had been played as follows:

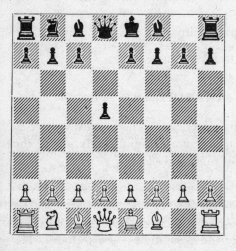

White	Black
N–KB3	P–Q4
N–K5	N–KB3
N–QB6	KN–Q2
N takes N	N takes N

FIGURE 50

Chessboard after Black's fourth move

An hour later, after losing a game to another player, I came back to see what Black and White were up to. In their second game the board looked exactly the same as it had before except that now *all four* knights were missing! Mr. Black, playing black, looked up and said, "I've just completed my *fifth* move."

a. Can the reader construct a legitimate game that leads to such a peculiar opening position?

"By the way," said Mr. White, "I've invented a problem that might amuse your readers. Suppose we dump a complete set of chessmen into a hat—all 16 black pieces and all 16 white— shake the hat, then remove the pieces randomly by pairs. If both are black, we put them on the table to form a black pile. If both happen to be white, we put them on the table to form a white pile. If the two pieces fail to match in color, we toss them into their chess box. After all 32 pieces have been removed from the hat, what's the probability that the number of pieces in the black pile will be exactly the same as the number in the white pile?"

"H'm," I said. "Offhand I'd guess the probability would be rather low." Black and White continued their game with subdued chuckles.

b. What is the exact probability that the two piles will be equal?

2. TALKATIVE EVE

THIS CRYPTARITHM (or alphametic, as some puzzlists prefer to call them) is an old one of unknown origin, surely one of the best and, I hope, unfamiliar to most readers:

$$\frac{\text{EVE}}{\text{DID}} = . \text{TALKTALKTALK.} \ . \ . \ .$$

The same letters stand for the same digits, zero included. The fraction EVE/DID has been reduced to its lowest terms. Its decimal form has a repeating period of four digits. The solution is unique. To solve it, recall that the standard way to obtain the simplest fraction equivalent to a decimal of n repeating digits is to put the repeating period over n 9's and reduce the fraction to its lowest terms.

3. THREE SQUARES

USING only elementary geometry (not even trigonometry), prove that angle C in Figure 51 equals the sum of angles A and B.

I am grateful to Lyber Katz for this charmingly simple problem. He writes that as a child he went to school in Moscow, where the problem was given to his fourth-grade geometry class for extra credit to those who solved it. "The number of blind alleys the problem leads to," he adds, "is extraordinary."

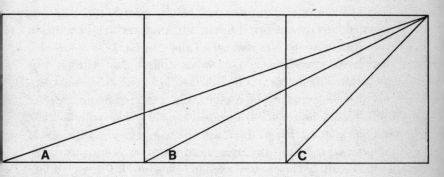

FIGURE 51
Prove that angle A *plus angle* B *equals angle* C.

4. POHL'S PROPOSITION

FREDERIK POHL, a top writer of science fiction, thought of this stunt, which appeared in a recent issue of a magic magazine called *Epilogue*. Computer programmers are likely to solve it more quickly than others.

Ask someone to draw a horizontal row of small circles on a sheet of paper to indicate a row of coins. Your back is turned while he does this. He then places the tip of his right thumb on the first circle so that his thumb and hand completely cover the row of circles. You turn around and bet you can immediately put on the sheet a number that will indicate the total number of combinations of heads and tails that are possible if each coin is flipped. For example, two coins can fall in four different ways, three coins in eight different ways and so on.

You have no way of knowing how many coins he drew and yet you win the bet easily. How?

5. ESCOTT'S SLIDING BLOCKS

THIS REMARKABLE SLIDING-BLOCK PUZZLE [*see Figure 52*] was invented by Edward Brind Escott, an American mathematician who died in 1946. It appeared in the August 1938 issue of a short-lived magazine called *Games Digest*. No solution was published. The problem is to slide the blocks one at a time, keeping them flat on the plane and inside the rectangular border, until block 1 and block 2 have exchanged places with block 7 and block 10, so that at the finish the two pairs of blocks are in the position shown at the right, with the other pieces anywhere on the board. No block is allowed to rotate, even if there is space for it to do so; each must keep its original orientation as it moves up, down, left or right.

It is the most difficult sliding-block puzzle I have seen in

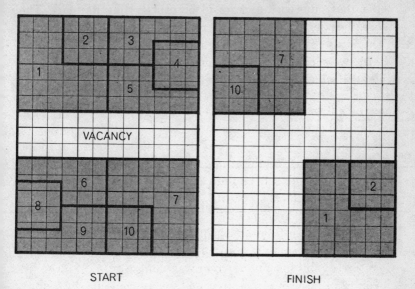

START   FINISH

FIGURE 52
Escott's sliding-block puzzle

print. The solution will be in 48 moves, counting each movement of one block as a single move even if it goes around a corner.

Escott was an expert on number theory who contributed frequently to mathematical journals. He taught at several schools and colleges in the Middle West and in his later years was an insurance company actuary in Oak Park, Ill.

6. RED, WHITE, AND BLUE WEIGHTS

PROBLEMS INVOLVING weights and balance scales have been popular during the past few decades. Here is an unusual one invented by Paul Curry, who is well known in conjuring circles as an amateur magician.

You have six weights. One pair is red, one pair white, one pair blue. In each pair one weight is a trifle heavier than the other but otherwise appears to be exactly like its mate. The three heavier weights (one of each color) all weigh the same. This is also true of the three lighter weights.

In two separate weighings on a balance scale, how can you identify which is the heavier weight of each pair?

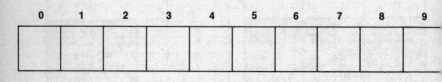

FIGURE 53
A digital problem

·7. THE 10-DIGIT NUMBER

IN THE 10 CELLS of Figure 53 inscribe a 10-digit number such that the digit in the first cell indicates the total number of zeros in the entire number, the digit in the cell marked "1" indicates the total number of 1's in the number, and so on to the last cell, whose digit indicates the total number of 9's in the number. (Zero is a digit, of course.) The answer is unique.

8. BOWLING-BALL PENNIES

KOBON FUJIMURA, the leading puzzle authority of Japan, devised this tricky little puzzle, which appears in one of his recent books. Arrange 10 pennies in the familiar bowling-pin formation [*see Figure 54*]. What is the smallest number of coins you must remove so that no equilateral triangle, of any size, will have its three corners marked by the centers of three pennies that remain? Not counting rotations and reflections as different, there is only one pattern for the removal of the minimum number of pennies.

FIGURE 54
Coin puzzle from Japan

Note that the pattern contains two equilateral triangles that are tipped so that their bases are not horizontal.

9. KNOCKOUT GEOGRAPHY

KNOCKOUT OPEN-END GEOGRAPHY is a word game for any number of players. The first player names any one of the 50 states. The next player must name a different state that either ends with the initial letter of the preceding state or begins with the last letter of the preceding state. For instance, if the first player gives Nevada, the next player can either affix Alaska or prefix Wisconsin. In other words, the chain of states remains open at both ends. When a player is unable to add to the chain, he is eliminated and the next player starts a new chain with a new state. No state can be named more than once in the same game. The game continues until only the winner remains.

David Silverman asks: If you are the first to name a state in a three-player game, what state can you name that will guarantee your winning? We assume that all players play rationally and without collusion to trap the first player.

ANSWERS

1. *a.* One possible line of play:

White	Black
1. N—KB3	N—KB3
2. N—QB3	N—QB3
3. N—Q4	N—Q4
4. KN takes N	QP takes N
5. N takes N	P takes N

Both chess problems were reprinted in the Summer 1969 issue of a mathematics magazine called *Manifold*, published at the University of Warwick in Coventry, England. It cited a 1947 issue of *Chess Review* as its source. The above variation is credited to Larry Blustein, an American player.

Mannis Charosh called my attention to an interesting variant of the problem of the two missing knights. Instead of removing the two king-side knights, remove the two queen-side knights, and instead of advancing the black queen's pawn two squares, advance it only one square. This too has a four-move solution, but it has the merit of being unique. (In the version I gave, Black's first two moves are interchangeable.) The problem appeared in the *Fairy Chess Review*, February 1955, where it was credited to G. Schweig, who had first published it in 1938. I leave the finding of the solution to the reader.

b. The probability is 1. Because the discarded set of pairs must contain an equal number of black and white pieces, the all-black and the all-white pile must be equal.

2. As stated earlier, to obtain the simplest fraction equal to a decimal of *n* repeated digits, put the repeating period over *n* 9's and reduce to its lowest terms. In this instance TALK/9,999, reduced to its lowest terms, must equal EVE/DID. DID, consequently, is a factor of 9,999. Only three such factors fit DID: 101, 303, 909.

If DID = 101, then EVE/101 = TALK/9,999, and EVE = TALK/

99. Rearranging terms, TALK = (99) (EVE). EVE cannot be 101 (since we have assumed 101 to be DID) and anything larger than 101, when multiplied by 99, has a five-digit product. And so DID = 101 is ruled out.

If DID = 909, then EVE/909 = TALK/9,999, and EVE = TALK/ 11. Rearranging terms, TALK = (11) (EVE). In that case the last digit of TALK would have to be E. Since it is not E, 909 also is ruled out.

Only 303 remains as a possibility for DID. Because.EVE must be smaller than 303, E is 1 or 2. Of the 14 possibilities (121, 141, ..., 292) only 242 produces a decimal fitting .TALK-TALK..., in which all the digits differ from those in EVE and DID.

The unique answer is 242/303 = .798679867986.... If EVE/ DID is not assumed to be in lowest terms, there is one other solution, 212/606 = .349834983498 ..., proving, as Joseph Madachy has remarked, that EVE double-talked.

3. There are many ways to prove that angle C in the figure is the sum of angles A and B. Here is one [*see Figure 55*]. Construct the squares indicated by gray lines. Angle B equals angle

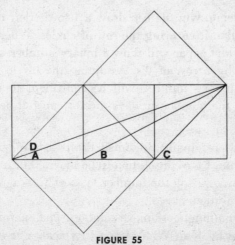

FIGURE 55
Construction for proof of three-square theorem

D because they are corresponding angles of similar right triangles. Since angles A and D add to angle C, B can be substituted for D, and it follows immediately that C is the sum of A and B.

This little problem produced a flood of letters from readers who sent dozens of other proofs. Scores of correspondents avoided construction lines by making the diagonals equal to the square roots of 2, 5, and 10, then using ratios to find two similar triangles from which the desired proof would follow. Others generalized the problem in unusual ways.

Charles Trigg published 54 different proofs in the *Journal of Recreational Mathematics*, Vol. 4, April 1971, pages 90–99. A proof using paper cutting, by Ali R. Amir-Moéz, appeared in the same journal, Vol. 5, Winter 1973, pages 8–9. For other proofs, see Roger North's contribution to *The Mathematical Gazette*, December 1973, pages 334–36, and its continuation in the same journal, October 1974, pages 212–15. For a generalization of the problem to a row of n squares, see Trigg's "Geometrical Proof of a Result of Lehmer's," in *The Fibonacci Quarterly*, Vol. 11, December 1973, pages 539–40.

4. In order to win the bet, draw a 1 to the left of the tip of the thumb that is covering the row of circles. When the thumb is removed, the paper will show a binary number consisting of 1 followed by a row of 0's. Assuming the 0's to represent n coins, this binary number will be equivalent to the decimal number 2^n, the number of ways n coins can fall heads or tails.

5. When I originally presented Escott's sliding-block puzzle in my column, I gave a solution in 66 moves, but many readers succeeded in lowering the number to 48. This is now the shortest known solution.

There is no unique 48-move solution. The one given in Figure 56 (sent by John W. Wright) is typical. The letters U, D,

Move	Block	Direction		Move	Block	Direction
1	6	U, R		25	4	U
2	1	D		26	2	U
3	5	L		27	10	U
4	6	L		28	9	R
5	4	D		29	7	U
6	5	R		30	1	U
7	2	D		31	8	U, R
8	3	L		32	1	D
9	5	U		33	7	L
10	2	R		34	10	L
11	6	U, L		35	2	L
12	4	L, U		36	4	D
13	7	U		37	2	R, D
14	10	R		38	3	D
15	9	R		39	6	R
16	8	D		40	5	R
17	1	D		41	7	U
18	7	L		42	10	L, U, L, U
19	2	D		43	8	U, L, U, L
20	4	R, D		44	4	L, U
21	5	D		45	2	D, L
22	3	R		46	9	U
23	6	U		47	2	R
24	5	L, U		48	1	R

FIGURE 56

A 48-move solution to sliding-block puzzle

L, and *R* stand for up, down, left, and right. In each case the numbered piece moves as far as possible in the indicated direction.

Since the initial pattern has twofold symmetry, every solution has its inverse. In this case the inverse starts with piece 5 moving down and to the left instead of piece 6 moving up and to the right and continues with symmetrically corresponding moves.

6. One way to solve the problem of the six weights—two red, two white, and two blue—is first to balance a red and a white weight against a blue and a white weight.

If the scales balance, you know there are a heavy and a light weight on each pan. Remove both colored weights, leaving the white weights, one on each side. This establishes which white weight is the heavier. At the same time it tells you which of the other two weights used before (one red, one blue) is heavy and which is light. This in turn tells you which is heavy and which is light in the red-blue pair not yet used.

If the scales do not balance on the first weighing, you know that the white weight on the side that went down must be the heavier of the two whites, but you are still in the dark about the red and blue. Weigh the original red against the *mate* of the original blue (or the original blue against the *mate* of the original red). As C. B. Chandler (who sent this simple solution) put it, the result of the second weighing, plus the memory of which side was heavier in the first weighing, is now sufficient to identify the six weights.

For readers who liked working on this problem, Ben Braude, a New York City dentist and amateur magician, devised the following variation. The six weights are alike in all respects (including color) except that three are heavy and three light. The heavy weights weigh the same and the light weights weigh the same. Identify each in three separate weighings on a balance scale.

As Thomas O'Beirne pointed out, Braude's problem has two essentially different types of solutions: one in which pairs are weighed against pairs, and one in which each weighing involves a single weight on each side. John Hamilton gave this concise chart for the four possibilities of the simpler method. (It appeared in the March 1970 issue of the magic periodical, *The Pallbearers Review*.)

1	2	3	4
a\B	a\B	a—b	a—b
c\D	c—d	b—c	b\C
e\F	d\E	c\D	D—E

A capital letter indicates a heavy weight, a small letter a light weight. A horizontal line means balance, a slanted line shows how the scale tips.

7. The only answer is 6,210,001,000. I do not have the space for a detailed proof, but a good one by Edward P. DeLorenzo is in Allan J. Gottlieb's puzzle column in the Massachusetts Institute of Technology's *Technical Review* for February 1968. The same column for June 1968 has a proof by Kenneth W. Dritz that for fewer than 10 cells the only answers in the decimal system are 1,210; 2,020; 21,200; 3,211,000; 42,101,000, and 521,001,000.

See *Journal of Recreational Mathematics*, Vol. 11, 1978–79, pages 76–77, for Frank Rubin's general solution. He shows that no "tally number" exists for bases 1, 2, 3, and 6. For base 5 there is only 521,200. For base 4 there are 1,210 and 2,020. It is the only base with more than one solution. For all bases above 6 there is one solution, of the form $R21(0 \ldots 0)1000$, where R is 4 less than the base, and the number of zeroes inside the parentheses is 7 less than the base.

8. The four pennies shown shaded [*see Figure 57*] are the fewest that must be removed from the 10 so that no three re-

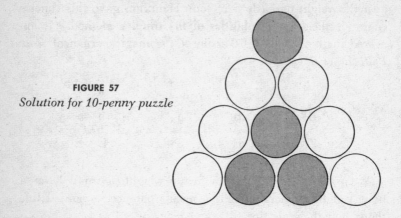

fewest that must be removed from the 10 so that no three remaining coins mark the corners of an equilateral triangle. Barring rotations, the pattern is unique; it is, of course, identical with its reflection.

9. A simple way to win David Silverman's geography game is to name Tennessee. The second player can only prefix Connecticut or Vermont. Since no state begins with E or ends with C or V, the third player is eliminated. It is now your turn to start again. You can win with Maine or Kentucky. Maine eliminates the second player immediately because no state begins with E or ends with M. Kentucky is a quick winner, among several other possibilities. It forces him to name New York. You win by prefixing Michigan, Washington, or Wisconsin.

Three other first moves also win for the first player on his second move: Delaware, Rhode Island, and Maryland. Other states, such as Vermont, Texas, and Connecticut, lead to wins on the first player's third move.

CHAPTER 12

Dominoes

SURPRISINGLY LITTLE seems to be known about the early history of dominoes. There are no references to them in Western literature before the middle of the 18th century, when domino games were first played in Italy and France. Later they spread over Europe and to England and America. The standard Western set of dominoes has always consisted of 28 tiles that display all possible pairs of digits from 0 through 6 [*see Figure 58*]. Each digit appears eight times in the set. Larger sets that run from the "doublet" 0–0 (two blank squares) to 9–9 (55 tiles in all) or to 12–12 (91 tiles) have occasionally been sold to accommodate larger numbers of players. The tiles are usually black with sunken white spots. They may have been called dominoes because of the resemblance of the 1–1 tile to the black domino half-mask worn in masquerades.

No one knows whether European dominoes were invented independently or copied from the Chinese. In any case they were popular in China for centuries before they became known in Europe. Chinese dominoes, called *kwat p'ai*, have no blanks.

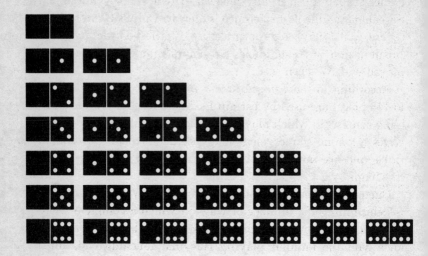

FIGURE 58
The standard set of 28 Western dominoes

A set includes every pair combination from 1–1 through 6–6 (21 tiles), but 11 tiles are duplicated, making 32 in all. As on Chinese dice, the ace and the 4 have red spots. All other spots are white (or black if the tiles are white) except on the 6–6, where three spots on one side of each 6 are red. (Korean dominoes are the same except that the ace is much larger than the other spots.) Each tile has a colorful Chinese name: 6–6 is "heaven," 1–1 is "earth," 5–5 is "plum flower," 6–5 is "tiger's head," and so on. The names are the same as those given to the corresponding 21 throws of a pair of dice.

Chinese dominoes are frequently made of cardboard rather than of such material as wood, ivory, and ebony and are then handled like playing cards. As in the West, many different games are played with the pieces. The most popular Western games are described in any modern "Hoyle." For Chinese and Korean games the best reference is Stewart Culin's *Games of the Orient*, an 1895 book reprinted by Charles Tuttle in 1958. Japan has no native dominoes but domino games are sometimes played with Western sets.

According to the *Encyclopaedia Britannica*, bone sets of 60 to 148 pieces are used by certain Eskimo tribes for frantic gambling contests in which players occasionally stake and lose their wives. Domino games have long been a favorite pastime in Cuba, and are now the primary recreation of Cuban refugees in Miami.

One of the oldest combinatorial problems involving dominoes is determining the number of ways in which a complete set of Western dominoes can be arranged in a straight row in accordance with the familiar playing rule that touching ends must match. (A set is complete if it contains all pairs from 0–0 through n–n.) The problem is interesting because it translates directly into a problem of graph-tracing [*see Figure 59*]. Ignoring the trivial set of one domino, 0–0, consider the simplest complete set: 0–0, 0–1, 1–1 [*a*]. The line from 0 to 1 corresponds to the 0–1 tile. The circles, showing that each digit is paired with itself, indicate the doublets in the set. The number of ways the three tiles can be arranged in a row is the same as the number of ways the simple graph can be traversed by a single path that does not go over any line twice. Obviously there are only two such paths, one a reverse of the other. These two (0–0, 0–1, 1–1, and its reversal) are the only two ways the tiles can be placed in a row with touching ends matching.

The problem is less trivial with the next-largest set of six tiles from 0–0 through 2–2. Its triangular graph [*b*] also has a unique path (and its reversal), but now the path must return to its starting spot. This means that the corresponding chain of dominoes is a closed ring: 0–0, 0–1, 1–1, 1–2, 2–2, 2–0. Think of the two ends as joined: 2–0, 0–0. Since the ring can be broken at six places to form a row, there are six different solutions, or 12 if reversals are counted as different.

The complete set of 10 dominoes (0–0 through 3–3) takes an unexpected turn [c]. All four vertexes are odd, that is, an odd number of lines meet at each. (The center crossing point of the two diagonals is not considered a vertex.) An old graph-tracing rule, first given by Leonhard Euler in his famous analysis of the problem of traversing the seven bridges of Königsberg, is that a graph can be traced by one path, without going over any line twice, if and only if all vertexes are even, or if there are exactly two odd vertexes. In the first case the path is always closed, ending where it began. In the second case the path must begin at one odd vertex and end at the other. Since we have here four odd vertexes, there is no single path that will trace the entire graph and therefore no way the 10 dominoes can form a row. An equivalent impossibility proof is to note that every digit appears five times in the complete set. Because each digit must appear *within* the row an even number of times—a result of the end-matching rule—it must appear once at one end of the row. There are four digits, but a row has only two ends, and so a single row is impossible. The best we can do is to traverse the graph with two unjoined paths, which correspond to two separate rows of tiles. The end digits of the two rows obviously must be 0, 1, 2, 3.

The "complete graph" for five spots, with circles added to join each spot to itself, corresponds to the complete set of 15 tiles, 0–0 through 4–4 [d]. Since all vertexes are even, a closed path can be drawn. (As on all such graphs, crossing points inside the polygon are not vertexes.) Counting the number of such paths, each of which can be broken at 15 places to make an open chain, is a moderately complicated task. Henry Ernest Dudeney, answering this problem in his *Amusements in Mathematics* (Problem 283), points out that the pentagonal graph, aside from its circles, can be traversed in 264 ways, each of which gives a ring of dominoes. (For example, the path that

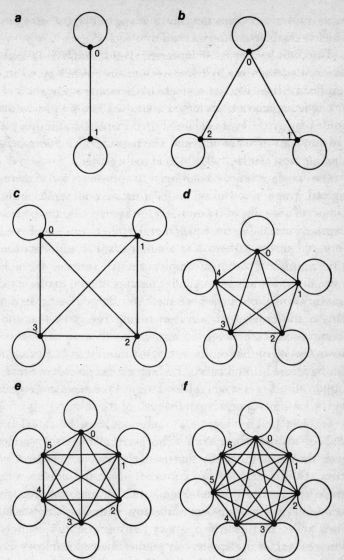

FIGURE 59
*Graphs for solving the row problem with
complete sets of dominoes*

starts 3024 . . . yields the ring that starts 3–0, 0–2, 2–4, . . .)
The five doublets can be inserted into each ring in $2^5 = 32$ ways,
making $264 \times 32 = 8{,}448$ different rings. Each ring can be
broken at 15 places; therefore we multiply 8,448 by 15 to get
126,720 different row arrangements, including reversals.

The hexagonal graph for six spots [e] has six odd vertexes.
Consequently the complete set of 21 dominoes, 0–0 through 5–5,
cannot be arranged in one row. The best we can achieve is three
separate rows with 0, 1, 2, 3, 4, 5 at the six ends.

The standard set of 28 dominoes, 0–0 through 6–6, has a hep-
tagonal graph [f]. Note that 28 is the second perfect number
(equal to the sum of its divisors). All perfect numbers are triangular (sums of successive integers 1, 2, 3 . . .), and it takes
only a glance at Figure 58 to see that every triangular number
is the number of tiles in a complete set. All vertexes on the hep-
tagonal graph are even, and therefore closed paths can be
drawn. It turns out that there are 7,959,229,931,520 such paths!
This is the number of ways, including reversals, that the 28
dominoes can be arranged in a row. For all complete sets, with
the exception of the set whose highest number is 1, a single row
can be formed if and only if the highest number is even. If the
highest number n is odd, at least $(n + 1)/2$ rows are required,
with all n digits appearing at the ends of the rows.

The fact that a chain of 28 dominoes must be closed is the
basis of an old parlor trick. The performer secretly removes
from the set any domino that is not a doublet. He leaves the
room while the other players arrange all the dominoes in a
single row. After this has been done the magician is able to
name the two end numbers of the row without seeing the tiles.
They will, of course, be the two numbers on the domino he
removed. (If he prefers, he can predict the two numbers in advance by writing them on a piece of paper that is folded and put
aside.) To repeat the trick he replaces the stolen domino in the
act of shuffling the tiles and palms a different one.

Many domino problems have been proposed in which a com-

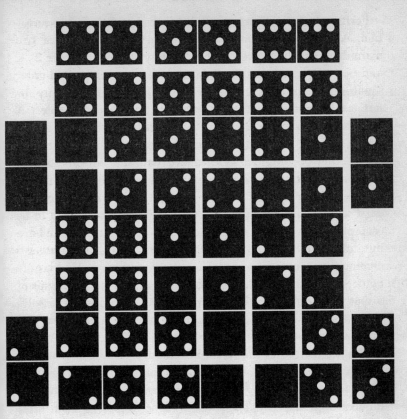

FIGURE 60

A sample quadrille

plete set is formed into a symmetrical polygon under certain restrictions. For example, the 19th-century French mathematician Edouard Lucas, in the second volume of his *Récréations Mathématiques*, introduced what are called "quadrilles," polygons in which the standard 28 tiles are so arranged that every digit forms two sets of two-by-two squares. A quadrille from Lucas's work is shown that has a unique solution except for permutations of the digits and reflections of the entire figure [*see Figure 60*].

Forming magic squares with dominoes is another old recreation. A square is magic if all rows and columns and the two main diagonals have the same sum. Only squares of order 2, 4, and 6 can be made with tiles from the set of 28. (Odd-order squares contain an odd number of cells, and therefore any attempt to form them with dominoes is bound to leave a hole.) A magic square of order 2 clearly is impossible; even if diagonals are disregarded, the two tiles would have to be duplicates.

An order-6 domino magic square with the lowest possible magic constant, 13 [*see Figure 61, top*], can be changed to a square with the highest possible constant, 23, by replacing each digit with its difference from 6. The two squares are said to be "complementary" with respect to 6. To prove that 13 and 23 are minimum and maximum constants, first note that an order-6 magic square must have a total of spots that is evenly divisible by 6. Since 78 and 138 are the smallest and largest multiples of 6 that can be the sum of the spots on 18 dominoes, it follows that $78/6 = 13$ and $138/6 = 23$ are the smallest and largest possible constants.

The smallest and largest constants for an order-4 magic square formed with eight dominoes from the standard set are $20/4 = 5$ and $76/4 = 19$. If one starts with a square with a constant of 5 [*at bottom left in Figure 61*], replacing each digit with its difference from 6 produces a magic square with the maximum constant, 19. Order-4 domino magic squares are possible with any constant from 5 through 19. Can the reader find eight dominoes from the standard set that will fit the blank pattern [*at bottom right in illustration*] to produce a magic square that adds to 10 along all rows and columns and the two main diagonals? In 1969 Wade E. Philpott proved that order-6 magic squares can be made with any constant from 13 through 23.

One may explore magic squares of odd order by adopting one of several inelegant conventions:

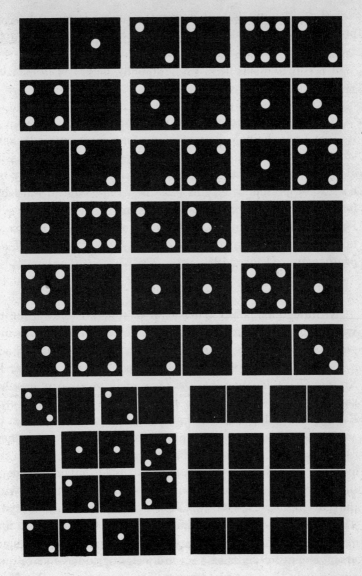

FIGURE 61
Domino magic squares

1. Leave a unit hole that counts as zero. It is not hard to prove that an order-3 magic square of this type is not possible.

2. Allow one cell of a tile, preferably blank, to project beyond the square.

3. Treat each domino as a single number that is the sum of its spots. Since single dominoes in a standard set of 28 have sums of 1 through 9, the unique order-3 magic square with digits 1 through 9 can be formed with nine dominoes. For orders 4 and 5, duplicate sums must be used. Leslie E. Card has found that any set of 25 dominoes from the standard set will form an order-5 magic square of this kind. (See "An Enumeration Problem" by David E. Silverman, *Journal of Recreational Mathematics*, October 1970, pages 226–27.)

An intriguing puzzle game with dominoes comes to me by way of Lech Pijanowski, a film critic in Warsaw who also writes a weekly newspaper column on games of mental skill and is the author of a 360-page book on board games, *Podroze W Krainie Gier* (*Journey into the Land of Games*). Any number can play but we shall assume that there are only two players. Each does as follows. While his opponent is out of the room he shuffles the 28 face-down tiles of a standard set, then forms them randomly into a seven-by-eight rectangle. The tiles are turned over and their digits copied on a grid without showing the domino pattern. (It is a good plan to make a second copy, showing the pattern, to prove later that such a pattern does indeed exist.) The patternless grids are exchanged and the first player to find a way of forming it with dominoes is the winner. Since many arrangements of digits on a seven-by-eight grid have more than one solution, it is not necessary to discover the original pattern —just a pattern that will produce the grid.

Given a patternless grid [*see "a" in Figure 62*], how does one go about finding a solution? Pijanowski suggests first listing

a

4	1	3	4	3	5	3	3
5	0	4	1	1	5	0	2
0	1	2	0	2	1	6	2
2	5	1	0	6	4	0	0
5	3	5	6	6	6	5	3
6	4	3	0	2	1	5	6
6	2	3	2	4	1	4	4

b

4	1	3	4	3	5	3	3
5	0	4	1	1	5	0	2
0	1	2	0	2	1	6	2
2	5	1	0	6	4	0	0
5	3	5	6	6	6	5	3
6	4	3	0	2	1	5	6
6	2	3	2	4	1	4	4

FIGURE 62

Solving a domino grid problem (see overleaf)

c

4	1	3	4	3	5	3	3
5	0	4	1	1	5	0	2
0	1	2	0	2	1	6	2
2	5	1	0	6	4	0	0
5	3	5	6	6	6	5	3
6	4	3	0	2	1	5	6
6	2	3	2	4	1	4	4

d

4	1	3	4	3	5	3	3
5	0	4	1	1	5	0	2
0	1	2	0	2	1	6	2
2	5	1	0	6	4	0	0
5	3	5	6	6	6	5	3
6	4	3	0	2	1	5	6
6	2	3	2	4	1	4	4

3	3	1	6	6	0	4
2	3	0	4	6	1	1
4	6	1	3	3	0	1
0	2	5	6	6	3	2
5	2	0	5	4	4	5
5	1	3	2	0	0	3
4	4	0	2	2	6	6

6	5	1	1	3	5	3	3
2	4	1	4	3	2	2	4
1	2	5	0	0	2	1	1
6	1	0	0	0	0	6	3
6	5	4	0	0	1	6	2
5	2	4	6	3	3	6	4
4	2	4	3	5	5	5	6

FIGURE 63
Two domino grid problems

all 28 domino pairs, then searching the grid for pairs that can be at only one spot. In this case 4–5, 2–2, 3–6, and 4–4 must be where they are shown in *b*. To prevent holes the 0–0 and 3–3 can immediately be added. Because this prevents the 0–0 and 3–3 from appearing elsewhere, the two small bars are drawn to show that a domino cannot cross either bar.

The 2–5 tile must be either horizontal or vertical as indicated by the broken lines [*c*]. In either case 0–1 must go in the spot shown, from which 1–3 and 0–4 follow to avoid duplicating 0–1. More bars can now be added. Proceeding in this way it is not hard to find a solution. Figure 62*d* shows one of four solutions.

The reader is urged to test his prowess on the slightly more difficult grid shown at the left in Figure 63. It has only one solution. If successful, the reader may feel sufficiently confident to tackle the extremely difficult grid shown at the right in the illustration. The two grids were supplied by Pijanowski. The second has eight solutions.

ANSWERS

Two of many solutions for the domino magic-square problem are given in Figure 64. The answer to the first domino grid problem is unique [*see left drawing in Figure 65*].

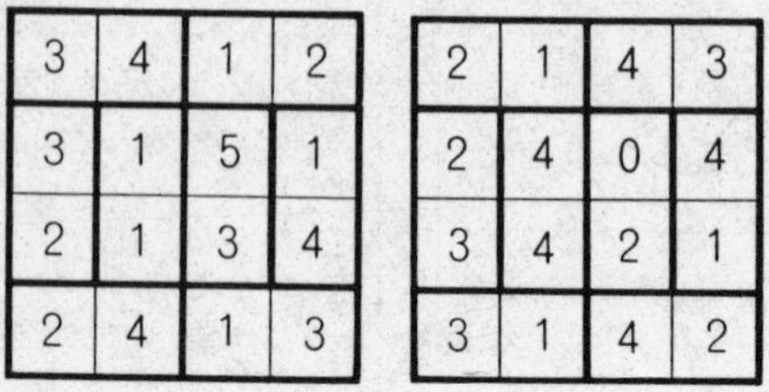

FIGURE 64
Magic-square solutions

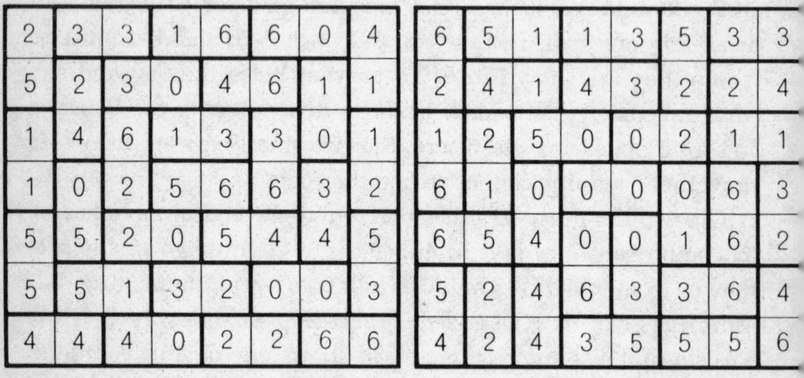

FIGURE 65
Domino grid solutions

The second grid problem has eight solutions. There are three basic patterns: the one shown at the right of Figure 65, which has a second form arising from a trivial rearrangement of the two shaded tiles; a second pattern also with two forms obtained by two arrangements of the same two tiles; and a third pattern in which there are two order-2 squares, each with two arrangements, to make four more solutions.

CHAPTER 13

Fibonacci and Lucas Numbers

Each wife of Fibonacci,
Eating nothing that wasn't starchy,
Weighed as much as the two before her.
His fifth was some signora!

—J. A. LINDON

THE GREATEST European mathematician of the Middle Ages was Leonardo of Pisa, better known as Fibonacci, meaning "son of Bonaccio" [*see Figure 66*]. Although Leonardo was born in Pisa, his father was an official of an Italian mercantile factory in Bougie in Algeria, and it was there that young Leonardo received his early mathematical training from Moslem tutors. He quickly recognized the enormous superiority of the Hindu-Arabic decimal system, with its positional notation and zero symbol, over the clumsy Roman system still used in his own country. His best-known work, *Liber abaci* (literally, "Book of the Abacus," but actually a comprehensive merchant's handbook on arithmetic and algebra), defended the merits of the Hindu-Arabic notation. The arguments made little impression on the Italian merchants of the time but the book eventually became the most influential single work in introducing the Hindu-Arabic system to the West. Although *Liber abaci* was

FIGURE 66
Fibonacci

completed in Pisa in 1202, it survives only in a revised 1228 edition dedicated to a famous court astrologer of the period. There has never been an English translation.

It is ironic that Leonardo, who made valuable contributions to mathematics, is remembered today mainly because a 19th-century French number theorist, Edouard Lucas (who edited a classic four-volume work on recreational mathematics), attached the name Fibonacci to a number sequence that appears in a trivial problem in *Liber abaci*. Suppose, Leonardo wrote, a male-female pair of adult rabbits is placed inside an enclosure to breed. Assume that rabbits start to bear young two months after their own birth, producing only a single male-female pair, and that they have one such pair at the end of each subsequent month. If none of the rabbits die, how many pairs of rabbits will there be inside the enclosure at the end of one year?

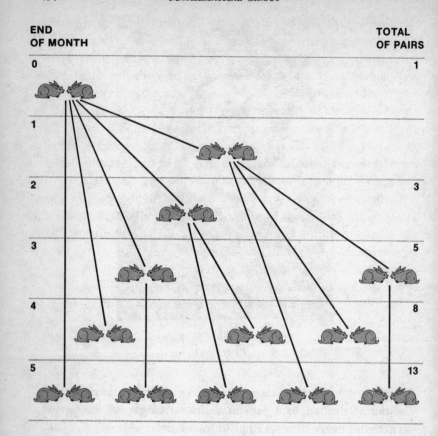

FIGURE 67
Tree graph for Fibonacci's rabbits

The tree graph [*see Figure 67*] shows what happens during the first five months. It is easy to see that the numbers of pairs at the close of each month form the sequence 1, 2, 3, 5, 8 . . . , in which each number (as Fibonacci pointed out) is the sum of the two numbers preceding it. At the end of 12 months there will be 377 pairs of rabbits.

Fibonacci did not investigate the sequence, and no serious study of it was undertaken until the beginning of the 19th century, when, as a mathematician once put it, papers on the sequence began to multiply almost as fast as Fibonacci's rabbits. Lucas made a deep study of sequences (now called "generalized Fibonacci sequences") that begin with *any* two positive integers, each number thereafter being the sum of the preceding two. He called the simplest such series, 1, 1, 2, 3, 5, 8, 13, 21 . . . , the Fibonacci sequence. (The next-simplest series, 1, 3, 4, 7, 11, 18 . . . , is now called the Lucas sequence.) The position of each number in this sequence is traditionally indicated by a subscript, so that $F_1 = 1$, $F_2 = 1$, $F_3 = 2$, and so on. (The first 50 Fibonacci numbers are listed in Figure 68). F_n refers to any Fibonacci number. F_{n+1} is the number following F_n; F_{n-1} is the number preceding F_n; F_{2n} is the F-number with a subscript twice that of F_n, and so on.

The Fibonacci sequence has intrigued mathematicians for centuries, partly because it has a way of turning up in unexpected places but mainly because the veriest amateur in number theory, with no knowledge beyond simple arithmetic, can explore the sequence and discover a seemingly endless variety of curious theorems. Recent developments in computer programming have reawakened interest in the series because it turns out to have useful applications in the sorting of data, information retrieval, the generation of random numbers, and even in rapid methods of approximating maxima and minima values of complicated functions for which derivatives are not known.

Early results are summarized in Chapter 17 of the first volume of Leonard Eugene Dickson's *History of the Theory of Numbers*. For the most recent discoveries interested readers can consult *The Fibonacci Quarterly*, published since 1963 by the Fibonacci Association. Edited by Verner E. Hoggatt, Jr., of San Jose State College in San Jose, Calif., the quarterly is concerned primarily with generalized Fibonacci numbers and similar numbers (such as "tribonacci numbers," which are sums of the preceding *three* numbers), but the journal is also devoted "to the study of integers with special properties."

F_1	1	L_1	1
F_2	1	L_2	3
F_3	2	L_3	4
F_4	3	L_4	7
F_5	5	L_5	11
F_6	8	L_6	18
F_7	13	L_7	29
F_8	21	L_8	47
F_9	34	L_9	76
F_{10}	55	L_{10}	123
F_{11}	89	L_{11}	199
F_{12}	144	L_{12}	322
F_{13}	233	L_{13}	521
F_{14}	377	L_{14}	843
F_{15}	610	L_{15}	1364
F_{16}	987	L_{16}	2207
F_{17}	1597	L_{17}	3571
F_{18}	2584	L_{18}	5778
F_{19}	4181	L_{19}	9349
F_{20}	6765	L_{20}	15127
F_{21}	10946	L_{21}	24476
F_{22}	17711	L_{22}	39603
F_{23}	28657	L_{23}	64079
F_{24}	46368	L_{24}	103682
F_{25}	75025	L_{25}	167761
F_{26}	121393	L_{26}	271443
F_{27}	196418	L_{27}	439204
F_{28}	317811	L_{28}	710647
F_{29}	514229	L_{29}	1149851
F_{30}	832040	L_{30}	1860498
F_{31}	1346269	L_{31}	3010349
F_{32}	2178309	L_{32}	4870847
F_{33}	3524578	L_{33}	7881196
F_{34}	5702887	L_{34}	12752043
F_{35}	9227465	L_{35}	20633239
F_{36}	14930352	L_{36}	33385282
F_{37}	24157817	L_{37}	54018521
F_{38}	39088169	L_{38}	87403803
F_{39}	63245986	L_{39}	141422324
F_{40}	102334155	L_{40}	228826127

FIGURE 68

The first 40 Fibonacci and Lucas numbers

Surely the most remarkable property of the Fibonacci series (which holds for the generalized series too) is that the ratio between two consecutive numbers is alternately greater or smaller than the golden ratio and that, as the series continues, the differences become less and less; the ratios approach the golden ratio as a limit. The golden ratio is a famous irrational number, 1.61803 . . . , that is obtained by halving the sum of 1 and the square root of 5. There is a considerable literature, some of it crankish, about the appearance of the golden ratio and the closely related Fibonacci sequence in organic growth and about their applications to art, architecture, and even poetry. George Eckel Duckworth, professor of classics at Princeton University, maintains in his book *Structural Patterns and Proportions in Vergil's Aeneid* (University of Michigan Press, 1962) that the Fibonacci series was consciously used by Vergil and other Roman poets of the time. I dealt with such matters in an earlier column on the golden ratio, which is reprinted in *The Second Scientific American Book of Mathematical Puzzles and Diversions.*

The most striking appearance of Fibonacci numbers in plants is in the spiral arrangement of seeds on the face of certain varieties of sunflower. There are two sets of logarithmic spirals, one set turning clockwise, the other counterclockwise, as indicated by the two shaded spirals in Figure 69. The numbers of spirals in the two sets are different and tend to be consecutive Fibonacci numbers. Sunflowers of average size usually have 34 and 55 spirals, but giant sunflowers have been developed that go as high as 89 and 144. In the letters department of *The Scientific Monthly* (November 1951), Daniel T. O'Connell, a geologist, and his wife reported having found on their Vermont farm one mammoth sunflower with 144 and 233 spirals!

The intimate connection between the Fibonacci series and the golden ratio can be seen in the following formula for the nth Fibonacci number:

$$F_n = \frac{1}{\sqrt{5}}\left[\left(\frac{1+\sqrt{5}}{2}\right)^n - \left(\frac{1-\sqrt{5}}{2}\right)^n\right].$$

This equation gives the nth Fibonacci number exactly (the $\sqrt{5}$'s cancel out), but it is cumbersome to use for high F-numbers, although good approximations can be obtained with loga-

FIGURE 69

Giant sunflower with 55 counterclockwise and 89 clockwise spirals

rithms. A much simpler formula for the nth F-number is the golden ratio raised to the power of n and then divided by the square root of 5. When this result is rounded off to the nearest integer, it too provides the exact number sought. Both formulas are nonrecursive because they compute the nth F-number directly from n. A "recursive procedure" is a series of steps each of which is dependent on previous steps. If you compute the nth F-number by summing consecutive F-numbers until you reach the nth, you are computing it recursively; a definition of the nth F-number as the sum of the preceding two numbers is a simple example of a recursive formula.

The formula that gives the nth number of the Lucas sequence exactly is

$$L_n = \left(\frac{1 + \sqrt{5}}{2}\right)^n + \left(\frac{1 - \sqrt{5}}{2}\right)^n,$$

but as in the case with Fibonacci numbers, there is a much simpler way to find the nth Lucas number. Simply raise the golden ratio to the power of n and round to the nearest integer.

Given any Fibonacci number greater than 1, you don't need to know its subscript to calculate the next Fibonacci number. Call the given number A. The next Fibonacci number is

$$\left[\frac{A + 1 + \sqrt{5A^2}}{2}\right],$$

where the brackets indicate rounding down to the nearest integer. The *same* formula gives the next Lucas number for any Lucas number greater than 3.

In a generalized Fibonacci sequence the sum of the first n terms is F_{n+2} minus the second term of the series. This is the basis of a pleasant lightning-calculation trick. Have someone put down any two starting numbers, then write down as many terms as he likes in a generalized Fibonacci sequence. Let him draw a line between any two numbers and you can quickly give the sum of all the terms up to the line. All you need do is note the second term past the line and subtract from it the second term of the sequence. If it is a standard Fibonacci sequence, you subtract 1; if it is a Lucas sequence, you subtract 3.

Here are some well-known properties of the standard Fibonacci sequence. Most of them are not difficult to prove, and of course all of them follow as special cases of theorems for the generalized sequence.

1. The square of any F-number differs by 1 from the product of the two F-numbers on each side. The difference is alternately plus or minus as the series continues. (For Lucas numbers the constant difference is 5.) See Chapter 8 of my *Mathematics, Magic and Mystery* for a famous geometrical dissection paradox in which this theorem plays a fundamental role. In a generalized Fibonacci sequence, the constant difference is $\pm (F_2^2 - F_1^2 - F_1 F_2)$.

2. The sum of the squares of any two consecutive F-numbers, F_n^2 and F_{n+1}^2, is F_{2n+1}. Since the last number must have an odd subscript, it follows from this theorem that if you write in sequence the squares of the F-numbers, sums of consecutive squares will produce in sequence the F-numbers with odd subscripts.

3. For any four consecutive F-numbers, A, B, C, D, the following formula holds: $C^2 - B^2 = A \times D$.

4. The sequence of final digits of the Fibonacci sequence repeats in cycles of 60. The last two digits repeat in cycles of 300. The repeating cycle is 1,500 for three final digits, 15,000 for four digits, 150,000 for five, and so on for all larger numbers of digits.

5. For every integer m there is an infinite number of F-numbers that are evenly divisible by m, and at least one can be found among the first $2m$ numbers of the Fibonacci sequence. This is not true of the Lucas sequence. No L-number, for instance, is a multiple of 5.

6. Every third F-number is divisible by 2, every fourth number by 3, every fifth number by 5, every sixth number by 8, and so on, the divisors being F-numbers in sequence. Consecutive Fibonacci numbers (as well as consecutive Lucas numbers) cannot have a common divisor other than 1.

7. With the exception of 3, every F-number that is prime has a prime subscript (for example, 233 is prime and its subscript, 13, is also prime). Put another way, if a subscript is composite (not prime), so is the number. Unfortunately the converse is not always true: a prime subscript does not necessarily mean that the number is prime. The first counterexample is F_{19}, 4,181. The subscript is prime but 4,181 is 37 times 113.

If the converse theorem held in all cases, it would answer the greatest unsolved question about Fibonacci numbers: Is there an infinity of Fibonacci primes? We know that the number of primes is infinite, and therefore if every F-number with a prime subscript were prime, there would be an infinity of prime F-numbers. As it is, no one today knows if there is a largest Fibonacci prime. The same question is also open for the Lucas sequence. The largest known F prime is F_{531}, a number of 119 digits. The largest known L prime is L_{353}, a number of 74 digits.

8. With the trivial exceptions of 0 and 1 (taking 0 to be F_0), the only square F-number is F_{12}, 144—which, surprisingly, is the square of its subscript. Whether or not there is a square F-number greater than 144 was an open question until the matter was finally settled, as recently as 1963, by John H. E. Cohn of Bedford College in the University of London. He also proved that 1 and 4 are the only squares in the Lucas sequence.

9. The only cubes among the Fibonacci numbers are 1 and 8, and the only cube among the Lucas numbers is 1. (See "On Fibonacci and Lucas Numbers Which Are Perfect Powers," Hymie London and Raphael Finkelstein, in *The Fibonacci Quarterly*, Vol. 77, December 1969, pages 476–81.)

10. The reciprocal of 89, the 11th F-number, can be generated by writing the Fibonacci sequence, starting with 0, and then adding as follows:

.0112358
13
21
34
55
89
144
233
337
610

. . .

. . . .

. . . .

. . . .
_____
.011235955056 . . . .

$1/89 = .011235955056179775 . .$

This list of properties could be extended to fill a book. One could do the same with instances of how the series applies to physical and mathematical situations. Leo Moser studied the paths of slanting light rays through two face-to-face glass plates. An unreflected ray goes through the plates in only one way [*see Figure 70*]. If a ray is reflected once, there are two paths; if it is reflected twice, there are three paths, and if three times, there are five paths. As n, the number of reflections, increases, the numbers of possible paths fall into the Fibonacci sequence. For n reflections the number of paths is F_{n+2}.

The sequence can be applied similarly to the different paths that can be taken by a bee crawling over hexagonal cells [*see Figure 71*]. The cells extend as far as desired to the right. Assume that the bee always moves to an adjacent cell and always moves toward the right. It is not hard to prove there is one path to cell 0, two paths to cell 1, three to cell 2, five to cell 3 and so on. As before, the number of paths is F_{n+2}, where n is the number of cells involved.

A male bee, or drone, by the way, has no father. As C. A. B. Smith has pointed out, a drone has 1 parent (its mother), 2 grandparents (its mother's parents), 3 great-grandparents (be-

**NUMBER
OF REFLECTIONS**

**NUMBER OF
DIFFERENT PATHS**

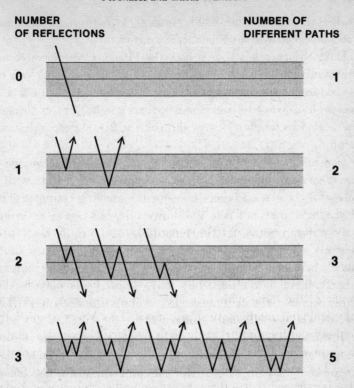

FIGURE 70

*There are F_{n+2} paths by which a ray can be reflected n times
through two panes of glass.*

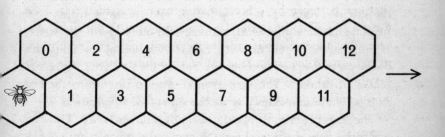

FIGURE 71

There are F_{n+2} paths by which the bee can crawl to cell n.

cause its mother's father had no father), 5 great-great-grandparents, and so on into the Fibonacci sequence.

David Klarner has shown how the Fibonacci numbers count the number of ways that dominoes (1×2 rectangles) can be packed into $2 \times k$ rectangles. There is 1 way to pack the 2×1 rectangle, 2 ways to pack the 2×2 rectangle, 3 ways to pack the 2×3 rectangle, 5 ways to pack the 2×4 rectangle, and so on.

Consider Fibonacci nim, a counter-removal game invented a few years ago by Robert E. Gaskell. The game begins with a pile of n counters. Players take turns removing counters. The first player may not take the entire pile, but thereafter either player may remove all the remaining counters if these rules permit: at least one counter must be taken on each play, but a player may never take more than twice the number of counters his opponent took on his last play. Thus if one player takes three counters, the next player may not take more than six. The person who takes the last counter wins.

It turns out that if n is a Fibonacci number, the second player can always win; otherwise the first player can win. If a game begins with 20 counters (not an F-number), how many must the first player take to be sure of winning?

A second problem concerns a little-known lightning-calculation trick. Turn your back and ask someone to write down any two positive integers (one below the other), add those two numbers to get a third, put the third number below the second, add the last two numbers to get a fourth, and continue in this way until he has a column of 10 numbers. In other words, he writes 10 numbers of a generalized Fibonacci sequence, each the sum of the preceding two numbers except for the first two, which are picked at random. You turn around, draw a line below the last number and immediately write the sum of all 10 numbers.

The secret is to multiply the seventh number by 11. This can easily be done in your head. Suppose the seventh number is 928. Put down the last digit, 8, as the last digit of the sum. Add

8 and 2 to get 10. Put 0 to the left of 8 in the sum, carrying the 1. The sum of the next pair of digits, 9 and 2, is 11. Add the carried 1 to get 12. Put 2 to the left of 0 in the sum, again carrying 1. Add the carried 1 to 9 and put down 10 to the left of 2 in the sum. The completed sum is 10,208. In brief, you sum the digits in pairs, moving to the left, carrying 1 when necessary, and ending with the last digit on the left.

Can you prove that the sum of the first 10 numbers in a generalized Fibonacci sequence is always 11 times the seventh number?

ADDENDUM

THE TRIBONACCI NUMBERS (1, 1, 2, 4, 7, 13, 24, 44, 81 . . .) were so-named by a brilliant young mathematician, Mark Feinberg, who contributed an article about them to *The Fibonacci Quarterly* (October 1963) when he was 14. His career at the University of Pennsylvania was cut short in 1967, his sophomore year, when he was killed in a motorcycle accident.

In his article on tribonacci numbers, Feinberg showed how the ratio between adjacent numbers, as the sequence grows, converges on .5436890126 . . . , the root of $x^3 + x^2 + x - 1 = 0$. We can generalize to sequences formed by summing four terms (tetranacci numbers), five terms, six terms, and so on. In all such sequences the ratio of adjacent terms converges on a limit. As the number of terms to be summed increases, the limiting ratio gets smaller, approaching .5 as a limit. This generalization had been published about 1913 by Mark Barr. (See my *Second Scientific American Book of Mathematical Puzzles and Diversions,* page 101.)

"Fibonacci notation" (introduced in the answer to the first problem), in which integers are uniquely expressed as sums of Fibonacci numbers, plays an important role in computer sorting techniques. See my *Scientific American* column for April 1973 for a way in which "Napier's abacus" (a little-known calculating device invented by the man who invented "Napier's bones") can be used for calculating in Fibonacci notation. On the role of Fibonacci notation in the strategy of playing Wythoff's game (a nim-like game), see my column for March 1977. For the appearance of the Fibonacci sequence in Pascal's triangle, see Chapter 15 of my *Mathematical Carnival*.

The Fibonacci and Lucas numbers are related by dozens of simple formulas. For examples: the nth Lucas number is equal to $F_{n-1} + F_{n+1}$. The product of F_n and L_n is equal to F_{2n}. The following Diophantine equation,

$$5x^2 \pm 4 = y^2,$$

has a solution in integers only when x is a Fibonacci number and y is the corresponding Lucas number.

The Fibonacci sequence and the Lucas sequence have in common the digits 1 and 3. Are there any larger numbers common to both sequences? The answer is no. See Martin D. Hirsch's note on "Additive Sequences," in *Mathematics Magazine*, Vol. 50, November 1977, page 264.

As pointed out earlier, the outstanding unsolved problem about the Fibonacci sequence is whether it contains a finite or infinite number of primes. In a generalized Fibonacci sequence, if the first two numbers are divisible by a prime, all its numbers are divisible by the same prime, and it is not hard to show that the sequence will contain a finite number of primes. If the first two numbers are coprime (have no common divisor), is there a generalized sequence that contains *no* primes?

This was first answered by R. L. Graham in "A Fibonacci-like Sequence of Composite Numbers," in *Mathematics Magazine*, Vol. 57, November 1964, pages 322–24. There are infinitely many such sequences, but the one with the smallest two starting numbers begins with:

1786772701928802632268715130455793,
1059683225053915111058165141686995.

ANSWERS

THE FIRST PROBLEM is to find the winning move in Fibonacci nim when the game begins with a pile of 20 counters. Since 25 is not a Fibonacci number, the first player has a sure win. To determine his first move he expresses 20 as the sum of Fibonacci numbers, starting with the largest possible F-number, 13, adding the next largest possible, 5, then the next, 2. Thus $20 = 13 + 5 + 2$. Every positive integer can be expressed as a unique sum in this way. No two consecutive F-numbers will appear in the expression. An F-number is expressed by one number only: itself.

The last number, 2, is the number of counters the first player must take to win. The second player is forbidden by the rules of Fibonacci nim to take more than twice 2, and therefore he cannot reduce the pile (which now has 18 counters) to the Fibonacci number 13. Assume that he takes four counters. The pile now contains 14. This is equal to F-numbers $13 + 1$, and so the first player takes one counter. By continuing this strategy he is sure to obtain the last counter and win.

If the initial number of counters is a Fibonacci number, say 144, the second player can always win. True, the first player can take 55 counters to leave 89, the next highest F-number, but then the second player can immediately win by taking all 89 counters because 89 is less than twice 55. The first player is forced, therefore, to leave a non-Fibonacci number of counters and the second player wins by the strategy that I have just explained. (See Donald E. Knuth, *Fundamental Algorithms*, Addison-Wesley, 1968, page 493, exercise No. 37, and "Fibonacci Nim," by Michael J. Whinihan in *The Fibonacci Quarterly*, Vol. 1, No. 4, December 1963, pages 9–13.)

To prove that the sum of the first 10 numbers in a generalized Fibonacci series is always 11 times the seventh number, call the

1.	a		
2.			b
3.	a	+	b
4.	a	+	2b
5.	2a	+	3b
6.	3a	+	5b
7.	5a	+	8b
8.	8a	+	13b
9.	13a	+	21b
10.	21a	+	34b
	55a	+	88b

FIGURE 72

Answer to Fibonacci problem

first two numbers *a* and *b*. The 10 numbers and their sum can be represented as shown in Figure 72. The sum obviously is 11 times the seventh number. Note that the coefficients of *a* and *b* form Fibonacci sequences.

CHAPTER 14

Simplicity

Fulfilling absolute decree
In casual simplicity.
— EMILY DICKINSON

Ms. DICKINSON'S LINES are about a small brown stone in a road, but if we view the stone as part of the universe, fulfilling nature's laws, all sorts of intricate and mysterious events are taking place within it on the microlevel. The concept of "simplicity," in both science and mathematics, raises a host of deep, complicated, still unanswered questions. Are the basic laws of nature few in number, or many, or perhaps infinite, as Stanislaw Ulam and others believe? Are the laws themselves simple or complex? Exactly what do we mean when we say one law or mathematical theorem is simpler than another? Is there any objective way to measure the simplicity of a law or a theory or a theorem?

Most biologists, especially those who are doing research on the brain, are impressed by the enormous complexity of living organisms. In contrast, although quantum theory has recently become more complicated with the discovery of hundreds of unexpected particles and interactions, most physicists retain a strong faith in the ultimate simplicity of basic physical laws.

This was especially true of Albert Einstein. "Our experience," he wrote, "justifies us in believing that nature is the realization of the simplest conceivable mathematical ideas." When he chose the tensor equations for his theory of gravitation, he picked the simplest set that would do the job, then published them with complete confidence that (as he once said to the mathematician John G. Kemeny) "God would not have passed up an opportunity to make nature that simple." It has even been argued that Einstein's great achievements were intellectual expressions of a psychological compulsion that Henry David Thoreau, in *Walden*, expressed as follows:

"Simplicity, simplicity, simplicity! I say, let your affairs be as two or three, and not a hundred or a thousand; instead of a million count half a dozen, and keep your accounts on your thumb nail."

In Peter Michelmore's biography of Einstein, he tells us that "Einstein's bedroom was monkish. There were no pictures on the wall, no carpet on the floor. . . . He shaved roughly with bar soap. He often went barefoot around the house. Only once every few months he would allow Elsa [his wife] to lop off swatches of his hair. . . . Most days he did not find underwear necessary. He also dispensed with pajamas and, later, with socks. 'What use are socks?' he asked. 'They only produce holes.' Elsa put her foot down when she saw him chopping off the sleeves of a new shirt from the elbow down. He explained that cuffs had to be buttoned or studded and washed frequently —all a waste of time."

"Every possession," Einstein said, "is a stone around the leg." The statement could have come straight out of *Walden*.

Yet nature seems to have a great many stones around her legs. Basic laws are simple only in first approximations; they become increasingly complex as they are refined to explain new observations. The guiding motto of the scientist, Alfred North Whitehead wrote, should be: "Seek simplicity and distrust it." Galileo picked the simplest workable equation for falling bodies, but it did not take into account the altitude of the body and had to be modified by the slightly more complicated equations of Newton. Newton too had great faith in simplicity.

"Nature is pleased with simplicity," he wrote, echoing a passage in Aristotle, "and affects not the pomp of superfluous causes." Yet Newton's equations in turn were modified by Einstein, and today there are physicists, such as Robert Dicke, who believe that Einstein's gravitational equations must be modified by still more complicated formulas.

It is dangerous to argue that because many basic laws are simple, the undiscovered laws also will be simple. Simple first approximations are obviously the easiest to discover first. Because the "aim of science is to seek the simplest explanations of complex facts," to quote Whitehead again (Chapter 7 of *The Concept of Nature*), we are apt to "fall into the error" of thinking that nature is fundamentally simple "because simplicity is the goal of our quest."

This we can say. Science sometimes simplifies things by producing theories that reduce to the same law phenomena previously considered unrelated—for example, the equivalence of inertia and gravity in general relativity. Science equally often discovers that behind apparently simple phenomena, such as the structure of matter, there lurks unsuspected complexity. Johannes Kepler struggled for years to defend the circular orbits of planets because the circle was the simplest closed curve. When he finally convinced himself that the orbits were ellipses, he wrote of the ellipse as "dung" he had to introduce to rid astronomy of vaster amounts of dung. It is a perceptive statement because it suggests that the introduction of more complexity on one level of a theory can introduce greater overall simplicity.

Nevertheless, at each step along the road simplicity seems to enter into a scientist's work in some mysterious way that makes the simplest workable hypothesis the best bet. "Simplest" is used here in a strictly objective sense, independent of human observation, even though no one yet knows how to define it. Naturally there are all sorts of ways one theory can be simpler than another in a pragmatic sense, but these ways are not relevant to the big question we are asking. As philosopher Nelson Goodman has put it, "If you want to go somewhere quickly, and several alternate routes are equally likely to be open, no one asks why you take the shortest." In other words, if two theories are not equivalent—lead to different predictions—and a scientist considers them equally likely to be true, he will test first the theory that he considers the "simplest" to test.

In this pragmatic sense, simplicity depends on a variety of factors: the kind of apparatus available, the extent of funding, the available time, the knowledge of the scientist and his assistants, and so on. Moreover, a theory may seem simple to one scientist because he understands the mathematics, and complicated to another scientist less familiar with the math. A theory may have a simple mathematical form but predict complex phenomena that are difficult to test, or it may be a complicated theory that predicts simple results. As Charles Peirce pointed out, circumstances may be such that it is more economical to test first the *least* plausible of several hypotheses.

These subjective and pragmatic factors obviously play roles in research, but they fail to touch the heart of the mystery. The deep question is: Why, other things being equal, is the simplest hypothesis usually the most likely to be on the right track—that is, most likely to be confirmed by future research?

Consider the following "simple" instance of a scientific investigation. A physicist, searching for a functional relationship

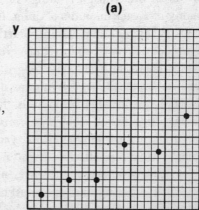

(a)

FIGURE 73
Observed data (*a*),
a possible function curve (*b*),
most likely function (*c*)

between two variables, records his observations as spots on a graph. Not only will he draw the simplest curve that comes close to the spots, he even allows simplicity to overrule the actual data. If the spots fall near a straight line, he will not draw a wavy curve that passes through every spot. He will assume that his observations are probably a bit off, pick a straight line that misses every spot, and guess that the function is a simple linear equation such as $x = 2y$ [*see Figure 73*]. If this fails to predict new observations, he will try a curve of next-higher degree, say a hyperbola or a parabola. The point is that, other things being equal, the simpler curve has the higher probability of being right. A truly astonishing number of basic laws are expressed by low-order equations. Nature's preference for extrema (maxima and minima) is another familiar example of simplicity because in both cases they are the values when the function's derivative equals zero.

Simplicity sometimes overrules data in evaluating even the most complex, high-level theories, such as the theory of relativity or theories about elementary particles. If a theory is sufficiently simple and beautiful, and has great explanatory power, these facts often count more for it than early experiments, which seem to falsify the theory, count against it.

This raises some of the most perplexing questions in the phi-

(b) **(c)**

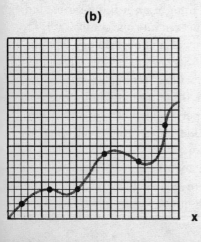 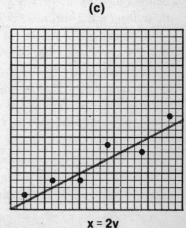

x = 2y

losophy of science. How can this particular kind of simplicity—the kind that adds to the probability that a law or theory is true—be defined? If it can be defined, can it be measured? Scientists tend to scorn both questions. They make intuitive judgments of simplicity without worrying about exactly what it is. Yet it is conceivable that someday a way to measure simplicity may have great practical value. Consider two theories that explain all known facts about fundamental particles. They are equal in their power to predict new observations, although the predictions are different. Both theories cannot be true. Both may be false. Each demands a different test and each test costs $1 million. If simplicity enters into the probability of a theory's being right, there is an obvious advantage in being able to measure simplicity so that the simplest theory can be tested first.

No one today knows how to measure this kind of simplicity or even how to define it. *Something* in the situation must be minimized, but what? It is no good to count the terms in a theory's mathematical formulation, because the number depends on the notation. The same formula may have 10 terms in one notation and three in another. Einstein's famous $E = mc^2$ looks simple only because each term is a shorthand symbol for concepts that can be written with formulas involving other concepts. This happens also in pure mathematics. The only way to express pi with integers is as the limit of an infinite series, but by writing π the entire series is squeezed into one symbol.

Minimizing the powers of terms also is misleading. For one thing, a linear equation such as $x = 2y$ graphs as a straight line only when the coordinates are Cartesian. With polar coordinates it graphs as a spiral. For another thing, minimizing powers is no help when equations are not polynomials. Even when they are polynomial, should one say that an equation such as $w = 13x + 23y + 132z$ is "simpler" than $x = y^2$?

In comparing the simplest geometric figures the notion of simplicity is annoyingly vague. In one of Johnny Hart's *B.C.* comic strips a caveman invents a square wagon wheel. Because it has too many corners and therefore too many bumps, he goes back to his drawing board and invents a "simpler" wheel in the shape of a triangle. Corners and bumps have been minimized, but the inventor is still further from the simplest wheel, the circle, which has no corners. Or should the circle be called the most complicated wheel because it is a "polygon" with an infinity of corners? The truth is that an equilateral triangle is simpler than a square in that it has fewer sides and corners. On the other hand, the square is simpler in that the formula for its area as a function of its side has fewer terms.

One of the most tempting of many proposed ways to measure the simplicity of a hypothesis is to count its number of primitive concepts. This, alas, is another blind alley. One can artificially reduce concepts by combining them. Nelson Goodman brings this out clearly in his famous "grue" paradox about which dozens of technical papers have been written. Consider a simple law: All emeralds are green. We now introduce the concept "grue." It is the property of being green if observed, say, before January 1, 2001, and being blue if observed thereafter. We state a second law: All emeralds are grue.

Both laws have the same number of concepts. Both have the same "empirical content" (they explain all observations). Both have equal predictive power. A single instance of a wrong color, when an emerald is examined at any future time, can falsify either hypothesis. Everyone prefers the first law because "green" is simpler than "grue"—it does not demand new theories to explain the sudden change of color of emeralds on January 1, 2001. Although Goodman has done more work than anyone on this narrow aspect of simplicity, he is still far from final results, to say nothing of the more difficult problem of measuring the overall simplicity of a law or theory. The concept of simplicity in science is far from simple! It may turn out that there is no single measure of simplicity but many different kinds, all of which enter into the complex final evaluation of a law or theory.

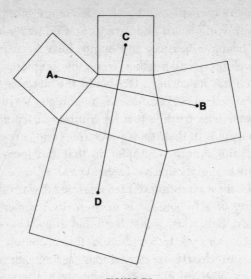

FIGURE 74
A "simple" geometrical theorem

Surprisingly, even in pure mathematics similar difficulties arise. Mathematicians usually search for theorems in a manner not much different from the way physicists search for laws. They make empirical tests. In pencil doodling with convex quadrilaterals—a way of experimenting with physical models—a geometer may find that when he draws squares outwardly on a quadrilateral's sides and joins the centers of opposite squares, the two lines are equal and intersect at 90 degrees [*see Figure 74*]. He tries it with quadrilaterals of different shapes, always getting the same results. Now he sniffs a theorem. Like a physicist, he picks the simplest hypothesis. He does not, for example, test first the conjecture that the two lines have a ratio of one to 1.00007 and intersect with angles of 89 and 91 degrees, even though this conjecture may equally well fit his crude measurements. He tests first the simpler guess that the lines are always perpendicular and equal. His "test," unlike the physicist's, is a search for a deductive proof that will establish the hypothesis with certainty.

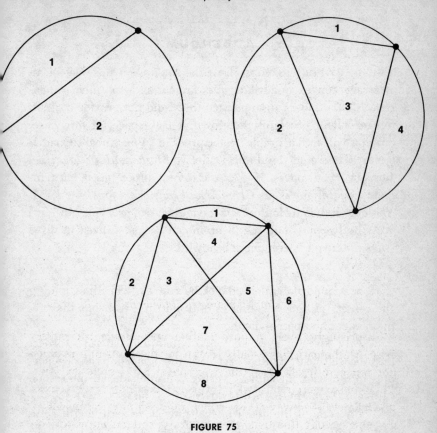

FIGURE 75
A combinatorial problem

Combinatorial theory is rich in similar instances where the simplest guess is usually the best bet. As in the physical world, however, there are surprises. Consider the following problem discovered by Leo Moser. Two or more spots are placed anywhere on a circle's circumference. Every pair is joined by a straight line. Given n spots, what is the maximum number of regions into which the circle can be divided? Figure 75 gives the answers for two, three, and four spots. The reader is asked to search for the answers for five and six spots and, if possible, find the general formula.

ADDENDUM

THE BEAUTIFUL THEOREM about the squares on the sides of an arbitrary convex quadrilateral is known as Von Aubel's theorem. Many readers, disappointed that I did not provide a proof, sent excellent proofs of their own. I lack space for any here, but you will find a simple vector proof in "Von Aubel's Quadrilateral Theorem," by Paul J. Kelly, *Mathematics Magazine*, January 1966, pages 35–37. A different proof based on symmetry operations is in *Geometric Transformations*, by I. M. Yaglom (Random House, 1962), pages 95–96, problem 24b.

As Kelly points out, the theorem can be generalized in three ways that make it even more beautiful.

1. The quadrilateral need not be convex. The lines joining the centers of opposite squares may not intersect, but they remain equal and perpendicular.

2. Any three or even all four of the quadrilateral's corners may be collinear. In the first case the quadrilateral degenerates into a triangle with a "vertex" on one side, in the second case into a straight line with two "vertices" on it.

3. One side of the quadrilateral may have zero length. This brings two corners together at a single point which may be treated as the center of a square of zero size.

The second and third generalizations were discovered by a reader, W. Nelson Goodwin, Jr., who drew the four examples shown in Figure 76. Note that the theorem continues to hold if opposite sides of a quadrilateral shrink to zero. The resulting line may be regarded as one of the lines connecting midpoints of opposite squares of zero size, and of course it equals and is perpendicular to a line joining midpoints of two squares drawn on opposite sides of the original line.

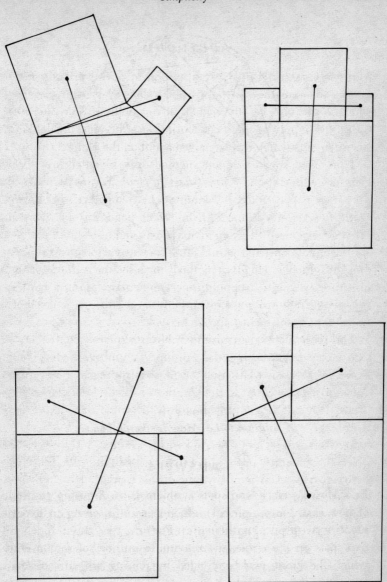

FIGURE 76
Curious generalizations of Von Aubel's theorem

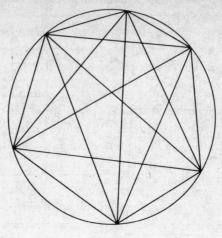

SPOTS								REGIONS
1				1				1
2			1		1			2
3			1	2	1			4
4		1	3		3	1		8
5		1	4	6	4	1		16
6	1	5	10	10	5	1		31
7	1	6	15	20	15	6	1	57

FIGURE 77

Solution to Leo Moser's spot problem

ANSWERS

LEO MOSER's circle-and-spots problem is an amusing example of how easily an empirical induction can go wrong in experimenting with pure mathematics. For one, two, three, four, and five spots on the circle, the maximum number of regions into which the circle can be divided by joining all pairs of spots with straight lines is 1, 2, 4, 8, 16. . . . One might conclude that this simple doubling series continues and that the maximum number of regions for n spots is 2^{n-1}. Unfortunately this formula fails for all subsequent numbers of spots. Figure 77 shows how six spots give a maximum of 31, not 32, regions. The correct formula is:

$$n + \binom{n}{4} + \binom{n-1}{2}.$$

A parenthetical expression $\binom{m}{k}$ is the number of ways m objects can be combined, taken k at a time. [It equals $m!/k!(m-k)!$] Moser has pointed out that the formula gives the sum of the rows of numbers at the left of the diagonal line drawn on Pascal's triangle, as shown in the illustration.

Written out in full, the formula is:

$$\frac{n^4 - 6n^3 + 23n^2 - 18n + 24}{24}.$$

When the positive integers are plugged into n, the formula generates the sequence: 1, 2, 4, 8, 16, 31, 57, 99, 163, 256, 386, 562. . . . The problem is a delightful illustration of Whitehead's advice to seek simplicity but distrust it.

I have been unable to determine where or when Moser first published this problem, but in a letter he says he thinks it was in *Mathematics Magazine* about 1950. It has since appeared in numerous books and periodicals, with varying methods of solution. A partial list of references is given in the bibliography for this chapter.

CHAPTER 15

The Rotating Round Table and Other Problems

1. ROTATING ROUND TABLE

In 1969, after 10 weeks of haggling, the Vietnam peace negotiators in Paris finally decided on the shape of the conference table: a circle seating 24 people, equally spaced. Assume that place cards on such a table bear 24 different names and that on one occasion there is such confusion that the 24 negotiators take seats at random. They discover that no one is seated correctly. Regardless of how they are seated, is it always possible to rotate the table until at least two people are simultaneously opposite their place cards?

A much more difficult problem arises if just one person finds his correct seat. Will it then always be possible to rotate the table to bring at least two people simultaneously opposite their cards?

2. SINGLE-CHECK CHESS

The British Chess Magazine, Vol. 36, No. 426, June 1916, reported that an American amateur named Frank Hopkins had invented a variation of chess, which he called "Single Check" or "One Check Wins." The game is played exactly like standard chess except that victory goes to the first player who checks (not checkmates) his opponent's king. The magazine, quoting from an article in *The Brooklyn Daily Eagle,* reported that "a suspicion that the white pieces had a sure win turned into a certainty" when U.S. grand master Frank J. Marshall one day "laconically remarked that he could 'bust' the new game." Hopkins was unconvinced until Marshall quickly won by moving *only his two white knights.* Marshall's strategy is not given, except for his opening move, nor does the account of the incident give the number of white moves before he administered the fatal check.

Since 1916 the idea of "single check" chess has occurred independently to many players. I first heard of it from Solomon W. Golomb, who knew it as "presto chess," a name given it by David L. Silverman after learning of the game from a 1965 reinventor. Back in the late 1940's the game had been reinvented by a group of mathematics graduate students at Princeton University. One of the students, William H. Mills, had discovered then what was undoubtedly Marshall's strategy: a way in which White, moving only his knights, can win on or before his fifth move. In 1969 Mills and George Soules together found different five-move wins involving pieces other than knights. Can you duplicate Marshall's feat by solving the problem shown in Figure 78? How can White, moving only his knights, check the black king in no more than five white moves?

FIGURE 78
*White to move knights
and check in five*

Attempts have been made to make the game more even by imposing additional rules. Hopkins himself proposed beginning with each player's pawns on the third row instead of the second. Sidney Sackson, who told me of the 1916 reference, suggests that the winner be the first to check a specified number of times, say from 5 to 10, depending on how long one wants the game to last. Whether either proposal effectively destroys White's advantage I cannot say.

3. WORD GUESSING GAME

ABOUT 1965 Anatol W. Holt, a mathematician who likes to invent new games, proposed the following word game. Two people each think of a "target word" with the same number of letters. Beginners should start with three-letter words and then go on to longer words as their skill improves. Players take turns calling out a "probe word" of the agreed length. The opponent must respond by saying whether the number of "hits" (right letter at the right position) is odd or even. The first to guess his opponent's word is the winner. To show how logical analysis can determine the word without guesswork, Holt has supplied the following example of six probe words given by one player:

Even	*Odd*
DAY	SAY
MAY	DUE
BUY	TEN

If you knew the target word and compared it letter by letter with any word on the even list, you would find that an even number of letters (zero counts as even) in each probe word would match letters at the same positions in the target word; words on the odd list would match the target word in an odd number of positions. Find the target word.

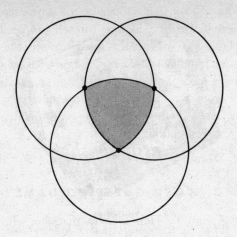

FIGURE 79
How much of one circle is shaded?

4. TRIPLE BEER RINGS

IN *Problematical Recreations*, No. 7, a series of puzzle booklets issued annually by Litton Industries in Beverly Hills, Calif., the following problem appeared. A man places his beer glass on the bar three times to produce the set of triple rings shown in Figure 79. He does it carefully, so that each circle passes through the center of the other two. The bartender thinks the area of mutual overlap (*shaded*) is less than one-fourth of the area of a circle, but the customer says it is more than one-fourth. Who is right?

The solution can be obtained the hard way by finding the area of an equilateral triangle inscribed in the shaded section and then adding the areas of the three segments of circles on each side of the triangle. A reader of this column, Tad Dunne of Willowdale in Ontario, sent me a beautiful graphical "look and see" solution that involves no geometrical formulas and almost no arithmetic, although it does make use of a repeating wallpaper pattern. Can the reader rediscover it?

FIGURE 80

What digits are on the cubes' hidden faces?

5. TWO-CUBE CALENDAR

IN GRAND CENTRAL TERMINAL in New York I saw in a store
window an unusual desk calendar [*see Figure 80*]. The day was
indicated simply by arranging the two cubes so that their front
faces gave the date. The face of each cube bore a single digit, 0
through 9, and one could arrange the cubes so that their front
faces indicated any date from 01, 02, 03 . . . to 31.

The reader should have little difficulty determining the four
digits that cannot be seen on the left cube and the three on the
right cube, although it is a bit trickier than one might expect.

6. UNCROSSED KNIGHT'S TOURS

IN THE *Journal of Recreational Mathematics* for July 1968,
L. D. Yarbrough introduced a new variant on the classic problem of the knight's tour. In addition to the rule that a knight
touring a chessboard cannot visit the same cell twice (except
for a final reentrant move that in certain tours allows the knight
to return to the starting square), the knight is also not permitted to cross its own path. (The path is taken to be a series of
straight lines joining the centers of the beginning and ending

squares of each leap.) The question naturally arises: What are the longest uncrossed knight's tours on square boards of various sizes?

Figure 81 reproduces samples of the longest uncrossed tours Yarbrough found for square boards from order-3 through order-

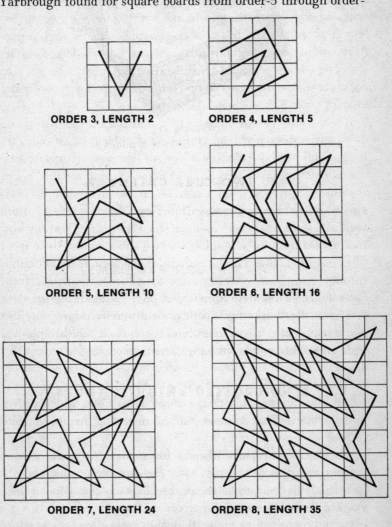

ORDER 3, LENGTH 2　　　　**ORDER 4, LENGTH 5**

ORDER 5, LENGTH 10　　　　**ORDER 6, LENGTH 16**

ORDER 7, LENGTH 24　　　　**ORDER 8, LENGTH 35**

FIGURE 81
Try to lengthen the order-6 knight's tour.

8. The order-7 tour is particularly interesting. Seldom does a re-entrant tour have maximum length; this is one of the rare exceptions as well as a tour of pleasing fourfold symmetry.

The idea of uncrossed knight's tours caught the interest of Donald E. Knuth. He wrote a "backtrack" computer program that, among other things, found every possible maximum-length uncrossed knight's tour on square boards through order-8. Rotations and reflections are, as usual, not considered different. The computer found two tours for the order-3 board, five for order-4, four for order-5, one for order-6, 14 for order-7, and four for order-8 (the standard chessboard).

It is the unique tour on the 6-by-6 that is most surprising and that provides our problem. Only on the order-6 board did Yarbrough fail to find a maximum-length uncrossed tour. His tour has 16 moves, but there is one uncrossed tour on this board that has 17 moves. The reader is invited to match his wits against the computer and see if he can discover the 17-move tour.

7. TWO URN PROBLEMS

PROBABILITY THEORISTS are fond of illustrating theorems with problems about identical objects of different colors that are taken from urns, boxes, bags, and so on. Even the simplest of such problems can be confusing. Consider, for instance, the fifth of Lewis Carroll's *Pillow Problems:* "A bag contains one counter, known to be either white or black. A white counter is put in, the bag shaken, and a counter drawn out, which proves to be white. What is now the chance of drawing a white counter?"

"At first sight," Carroll begins his answer, "it would appear that, as the state of the bag, *after* the operation, is necessarily identical with its state *before* it, the chance is just what it was, viz. 1/2. This, however, is an error."

Carroll goes on to show that the probability of a white

counter remaining in the bag actually is 2/3. His proof is a bit long-winded. Howard Ellis, a Chicago reader, has done it differently. Let B and $W(1)$ stand for the black or white counter that may be in the bag at the start and $W(2)$ for the added white counter. After removing a white counter there are three equally likely states:

In bag	Outside bag
$W(1)$	$W(2)$
$W(2)$	$W(1)$
B	$W(2)$

In two of these states a white counter remains in the bag, and so the chance of drawing a white counter the second time is 2/3.

A recent problem of this kind, with an even more surprising answer, begins with a bag containing an unknown number of black counters and an unknown number of white counters. (There must be at least one of each color.) The counters are taken out according to the following procedure. A counter is chosen at random and discarded. A second counter is taken at random. If it is the same color as before, it too is discarded. A third is chosen. If it matches the two previously taken counters, it is discarded. This continues, with the counters discarded as long as they match the color of the first counter.

Whenever a counter is taken that has a *different* color from the previous one, it is replaced in the bag, the bag is shaken, and the entire process begins again.

To make this crystal clear, here is a sample of how the first ten drawings might go:

1. First counter is black. Discard.
2. Next counter is black. Discard.
3. Next is white. Replace and begin again.
4. First is black. Discard.
5. Next is white. Replace and begin again.

6. First counter is white. Discard.

7. Next is white. Discard.

8. Next is black. Replace and begin again.

9. First counter is black. Discard.

10. Next is white. Replace and begin again.

It turns out, amazingly, that regardless of the ratio of white to black counters at the start, there is a fixed probability that the last counter left in the bag is black. What is that probability?

8. TEN QUICKIES

1. With a 7-minute hourglass and an 11-minute hourglass, what is the quickest way to time the boiling of an egg for 15 minutes? (From Karl Fulves.)

2. A man traveled 5,000 miles in a car with one spare tire. He rotated tires at intervals so that when the trip ended each tire had been used for the same number of miles. For how many miles was each tire used?

3. A standard deck of 52 cards is shuffled and cut and the cut is completed. The color of the top card is noted. This card is replaced on top, the deck is cut again, and the cut is completed. Once more the color of the top card is noted. What is the probability that the two cards noted are the same color?

4. Find a number base other than 10 in which 121 is a perfect square.

5. Draw six line segments of equal length to form eight equilateral triangles.

6. Assuming that the angle cannot be trisected by compass and straightedge, prove that no number in the doubling series 1, 2, 4, 8, 16, 32, . . . is a multiple of 3. (From Robert A. Weeks.)

7. A farmer has 20 pigs, 40 cows, 60 horses. How many horses does he have if you call the cows horses? (From T. H. O'Beirne.)

8. Translate: "He spoke from 2222222222222 people."

9. A Greek was born on the seventh day of 40 B.C. and died on the seventh day of A.D. 40. How many years did he live?

10. A woman either always answers truthfully, always answers falsely, or alternates true and false answers. How, in two questions, each answered by yes or no, can you determine whether she is a truther, a liar, or an alternater?

ANSWERS

1. IF A CIRCULAR table seats an even number of people, equally spaced, with place cards marking their spots, no matter how they seat themselves it is always possible to rotate the table until two or more are seated correctly. There are two initial situations to be considered:

a. No person sits correctly. The easy proof is based on what mathematicians call the "pigeonhole principle": If n objects are placed in $n - 1$ pigeonholes, at least one hole must contain two objects. If the table seats 24 people, and if every person is incorrectly seated, it clearly is possible to bring each person opposite his card by a suitable rotation of the table. There are 24 people but only 23 remaining positions for the table. Therefore at least two people must be simultaneously opposite their cards at one of the new positions. This proof applies regardless of whether the number of seats is odd or even.

b. One person sits correctly. Our task is to prove that the table can be rotated so that at least two people will be correctly seated. Proofs of this can be succinctly stated, but they are technical and require a knowledge of special notation. Here is a longer proof that is easier to understand. It is based on more than a dozen similar proofs that were supplied by readers.

Our strategy will be the *reductio ad absurdum* argument. We first assume it is *not* possible to rotate the table so that two persons are correctly seated, then show that this assumption leads to a contradiction.

If our assumption holds, at no position of the table will all persons be incorrectly seated because then the situation is the same as the one treated above, and which we disposed of by the pigeonhole principle. The table has 24 positions, and there are 24 people, therefore at each position of the table exactly one person must be correctly seated.

Assume that only Anderson is at his correct place. For each person there is a "displacement" number indicating how many places he is clockwise from his correct seat. Anderson's displacement is 0. One person will be displaced by 1 chair, another by 2 chairs, another by 3, and so on, to one person who is displaced by 23 chairs. Clearly no two people can have the same displacement number. If they did it would be possible to rotate the table to bring them both simultaneously to correct positions —a possibility ruled out by our assumption.

Consider Smith, who is improperly seated. We count chairs counterclockwise around the table until we get to Smith's place card. The count is equal, of course, to Smith's displacement number. Now consider Jones who is sitting where Smith is supposed to be. We continue counting counterclockwise until we come to Jones's place card. Again, the count equals Jones's displacement number. Opposite Jones's place card is Robinson. We count counterclockwise to Robinson's place card, and so on. Eventually, our count will return to Smith. If Smith and Jones had been in each other's seats, we would have returned to Smith after a two-person cycle that would have carried us just once around the table. If Smith, Jones, and Robinson are occupying one another's seats, we return to Smith after a cycle of three counts. The cycle may involve any number of people from 2 through 23 (it cannot include Anderson because he is correctly seated), but eventually the counting will return to where it started after circling the table an integral number of times. Thus the sum of all the displacement numbers in the cycle must equal 0 modulo 24—that is, the sum must be an exact multiple of 24.

If the cycle starting with Smith does not catch all 23 incorrectly seated persons, pick out another person not in his seat and go through the same procedure. As before, the counting must eventually return to where it started, after an integral number of times around the circle. Therefore the sum of the displacements in this cycle also is a multiple of 24. After one or more of such cycles, we will have counted every person's displacement. Since each cycle count is a multiple of 24, the sum of all the cycle counts must be a multiple of 24. In other words, we have shown that the sum of all the displacement numbers is a multiple of 24.

Now for the contradiction. The displacements are 0, 1, 2, 3 . . . 23. This sequence has a sum of 276, which is *not* a multiple of 24. The contradiction forces us to abandon our initial assumption and conclude that at least two people must have the same displacement number.

The proof generalizes to any table with an even number of chairs. The sum of $0 + 1 + 2 + 3 + \ldots + n$ is

$$\frac{n\,(n+1)}{2},$$

which is a multiple of n only when n is odd. Thus the proof fails for a table with an odd number of chairs.

George Rybicki solved the general problem this way. We start by assuming the contrary of what we wish to prove. Let n be the even number of persons, and let their names be replaced by the integers 0 to $n - 1$ "in such a way that the place cards are numbered in sequence around the table. If a delegate d originally sits down to a place card p, then the table must be rotated r steps before he is correctly seated, where $r = p - d$, unless this is negative, in which case $r = p - d + n$. The collection of values of d (and of p) for all delegates is clearly the integers 0 to $n - 1$, each taken once, but so also is the collection of values of r, or else two delegates would be correctly seated at the same time. Summing the above equations, one for each delegate, gives $S - S + nk$, where k is an integer and $S = n(n - 1)/2$, the sum of the integers from 0 to $n - 1$. It follows that $n = 2k + 1$, an odd number." This contradicts the original assumption.

"I actually solved this problem some years ago," Rybicki writes, "for a different but completely equivalent problem, a generalization of the nonattacking 'eight queens' problem for a cylindrical chessboard where diagonal attack is restricted to diagonals slanting in one direction only. I proved that this was insoluble for any board of even order. The above is the translation of my proof into the language of the table problem. Incidentally, the proof is somewhat easier if one is allowed to use congruences modulo n."

Donald E. Knuth also called attention to the equivalence of the round table and the queens problem, and cited an early solution by George Polya. Several readers pointed out that when the number of persons is odd, a simple arrangement that prevents more than one person from being seated correctly, regardless of how the table is rotated, is to seat them counterclockwise to their place-card order.

2. How can White, using only his knights, win a game of "single check" chess in five or fewer moves?

The opening move must be N (knight) to QB3. Since this threatens several different ways of checking in two moves, Black is forced to advance a pawn that will allow his king to move. If he advances the queen's pawn, N–N5 forces the black king to Q2, then N–KB3 leads to a check on White's fourth move. If Black moves his king-bishop pawn, N–N5 leads to a check on the third move. Black must therefore advance his king's pawn. If he advances it two squares, N–Q5 prevents the black king from moving and White wins on his third move. Black's only good response, therefore, is P–K3.

White's second move is N–K4. Black is forced to advance his king to K2. White's third move, N–KB3, can be met in many ways but none prevents a check on or before White's fifth move. If Black tries such moves as P–Q3, P–KB3, Q–K1, P–Q4, P–QB4, or N–KB3, White responds N–Q4 and wins on his next move. If Black tries P–K4, P–QB4, or N–QB3, White's N–KR4 wins on his next move.

In 1969 William H. Mills discovered that White could also check in five by opening N–QR3. Black must advance his king's pawn one or two squares. N–N5 forces Black's king to K2. White's third move, P–K4, is followed by White's Q–B3 or Q–R5, depending on Black's third move, and leads to a check on White's fifth move.

Two other opening moves leading to five-move checks have since been found by Mills and Georg Soules. They are P–K3 and P–K4. Against most of Black's replies Q–N4 leads to a three-move check. If Black's second move is N–KR3 or P–KR4, Q–B3 leads to a check in four. If Black's second move is P–K3, White moves Q–R5. Black must respond P–KN3. Then Q–K5 does the trick. (Black's Q–K2 is met by Q takes QBP; B–K2 is met by Q–N7; N–K2 is met by Q–B6.) If Black's second move is N–KB3, White's N–QR3 forces Black to advance his king's pawn one or two squares, then N–N5 forces Black to advance his king, and Q–B3 leads to a check on the next move.

David Silverman has suggested still another way to make single-check chess a playable game. The winner is the first to check with a piece that cannot be taken. So far as I know it is not known which player can always win if both sides play their best.

3. To determine the target word, label the six probe words as follows:

	Even		*Odd*
E1	DAY	*O1*	SAY
E2	MAY	*O2*	DUE
E3	BUY	*O3*	TEN

*E*1 and *E*2 show that the target word's first letter is not D or M, otherwise the parity (odd or even) would not be the same for both words. *E*1 and *O*1 show that the target word's first letter is either D or S, otherwise the parity could not be different for the two words. The first letter cannot be D and therefore must be S.

Since S is the first letter, *E*2 and *E*3 are wrong in their first letters. Both end in Y, therefore the second letter of the target word cannot be A or U, otherwise *E*2 and *E*3 could not have the same parity. Knowing that U is not the second letter and D not the first, *O*2 shows that E is the third letter. Knowing that the target word begins with S and ends with E, *O*3 shows that E is the second letter. The target word is SEE.

4. Three intersecting circles, each passing through the centers of the other two, can be repeated on the plane to form the wallpaper pattern shown in Figure 82. Each circle is made up of six delta-shaped figures (*D*) and 12 "bananas" (*B*). One-fourth of a circle's area must therefore equal the sum of one and a half deltas plus three bananas. The area common to three mutually intersecting circles (*shown shaded in the illustration*) consists of three bananas and one delta, and therefore it is smaller than one-fourth of a circle by an amount equal to half a delta. Computation shows that the mutual overlap is a little more than .22 of the circle's area.

5. Each cube must bear a 0, 1, and 2. This leaves only six faces for the remaining seven digits, but fortunately the same face can be used for 6 and 9, depending on how the cube is turned. The picture shows 3, 4, 5 on the right cube, and therefore its hidden faces must be 0, 1, and 2. On the left cube one can see 1 and 2, and so its hidden faces must be 0, 6 or 9, 7, and 8.

John S. Singleton wrote from England to say that he had patented the two-cube calendar in 1957/8 (British patent number

FIGURE 82
Solution to intersecting-circles problem

831572), but allowed the patent to lapse in 1965. For a variation of this problem, in which three cubes provide abbreviations for each month, see my *Scientific American* column for December 1977.

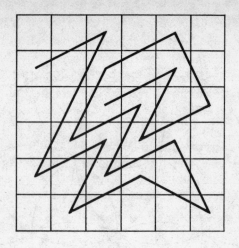

FIGURE 83
Maximum-length knight's tour on order-6 board

6. Figure 83 shows the unique maximum-length uncrossed knight's tour on the 6-by-6 board. For similar tours on higher-order square boards and on rectangular boards, see the *Journal of Recreational Mathematics*, Vol. 2, July 1969, pages 154–57.

7. It was stated that if counters are drawn according to a certain procedure from a bag containing an unknown mixture of white and black counters, there is a fixed probability, that the last counter will be black. If this is true, it must apply equally to each color. Therefore the probability is 1/2.

Although this answers the problem as posed, there remains the task of proving that the probability is indeed fixed. This can be done by induction, starting with two marbles and then going to three, four, and so on, or it can be done directly. Unfortunately both proofs are too long to give, so that I content myself with referring the reader to "A Sampling Process," by B. E. Oakley and R. L. Perry, in *The Mathematical Gazette* for February 1965, pages 42–44, where a direct proof is given.

One might hastily conclude that the solution generalizes; that is, if the bag contains a mixture of n colors, the probability that the last counter in the bag is a specified color is $1/n$. Unfortunately, this is not the case. As Perry pointed out in a letter, if there are 2 red, 1 white, and 1 blue counters, the probabilities that the last counter is red, white, or blue are, respectively, 26/72, 23/72, and 23/72.

8. The answers to the 10 quickies are as follows:

(1) Start the 7- and 11-minute hourglasses when the egg is dropped into the boiling water. When the sand stops running in the 7-glass, turn it over. When the sand stops in the 11-glass, turn the 7-glass again. When the sand stops again in the 7-glass, 15 minutes will have elapsed.

The above solution is the quickest one, but requires two turns of hourglasses. When the problem first appeared, I thoughtlessly asked for the "simplest" solution, having in mind the shortest one. Several dozen readers called attention to the following solution which is longer (22 minutes) but "simpler" in the sense of requiring only one turn. Start the hourglasses together. When the 7-minute one runs out, drop the egg in the boiling water. When the 11-minute hourglass runs out, turn it over. When it runs out a second time, the egg has boiled for 15 minutes.

If you enjoyed that problem, here is a slightly harder one of the same type, from Howard P. Dinesman's *Superior Mathematical Puzzles* (London: Allen and Unwin, 1968). What is the quickest way to measure 9 minutes with a 4-minute hourglass and a 7-minute hourglass?

(2) Each tire is used 4/5 of the total time. Therefore each tire has been used for 4/5 of 5,000 miles, or 4,000 miles.

(3) Whatever the color of the first card cut, this card cannot be the top card of the second cut. The second cut selects a card randomly from 51 cards of which 25 are the same color as the first card, and therefore the probability of the two cards' matching in color is 25/51, a bit less than 1/2.

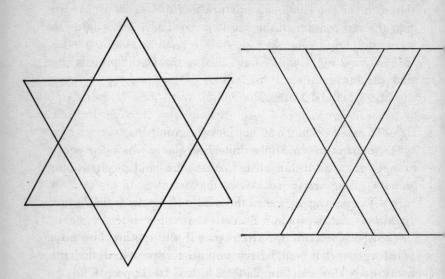

FIGURE 84
Six lines make eight triangles.

(4) 121 is a perfect square in any number notation with a base greater than 2. A quick proof is to observe that 11 times 11, in any system, has a product (in the same system) of 121. Craige Schensted showed that, with suitable definitions of "perfect square," 121 is a square even in systems based on negative numbers, fractions, irrational numbers, and complex numbers. "Although the bases may not be exhausted, I am and I assume you are, so I will stop here," he concluded.

(5) Figure 84 left shows how I answered the question. On the right is a second solution found independently by readers Harry Kemmerer and Gary Rieveschl.

(6) Any angle can be bisected with a compass and straightedge. By repeated bisections we can divide any angle into 2, 4, 8, 16, . . . equal parts. If any number in this series is a multi-

ple of 3, then repeated bisection obviously would allow trisection of the angle with compass and straightedge. Since this has been proved impossible, no number in the doubling series is evenly divisible by 3.

(7) The farmer has 60 horses. Calling a cow a horse doesn't make it a horse.

John Appel and Daniel Rosenblum were the first to tell me that this is a version of a joke attributed to Abraham Lincoln. He once asked a man who had been arguing that slavery was not slavery but a form of protection how many legs a dog would have if you called its tail a leg. The answer, said Lincoln, is four because calling a tail a leg does not make it a leg.

(8) The answer I gave to this was, "He spoke from 22 to 2 to 2:22 to 2,222 people." Readers sent other interpretations. David B. Eisendrath, Jr., wrote that if the speaker had been called a colonel it would have implied military time and the translation "He spoke from 22 to 22 to 22:22 to 22 people."

(9) The Greek lived 79 years. There was no year 0.

(10) Ask the woman "Are you an alternater?" twice. Two no answers prove she is a truther, two yes answers prove she is a liar, and a yes-no or no-yes response proves she is an alternater.

After the above answer appeared, several readers sent the following solution. I quote from a letter from Joseph C. Crowther, Jr.:

You can discover the inclination of the lady if you simply ask two questions about an obvious truth, such as "Do you have two ears?" or "Is water wet?" A truther will answer yes both times, and a liar will answer no. An alternater will not only say one of each, but the order in which she says them will determine which alternation she happens to be on, a fact which may prove useful in further conversation.

Ralph Seifert, Jr., sent a one-question solution that he attributed to his friend M. A. Zorn. "If someone asked you the same question twice, would you falsely answer 'no' exactly once?" The truther will say no, the liar yes, and the alternater will be so hopelessly befuddled that she won't be able to answer at all.

CHAPTER 16

Solar System Oddities

Around the ancient track marched, rank on rank,
The army of unalterable law.
                    —GEORGE MEREDITH, *Lucifer in Starlight*

ASTRONOMY, like every other science, has curious bypaths where one may stumble over mathematical problems with recreational aspects. In this chapter we take a quick look at the solar system, about which so many startling new discoveries are now being made, and consider some amusing mathematical questions that have arisen in the history of speculation about the structure of the sun's family of orbiting bodies.

First a bit of historical background. It is a common error to suppose all the ancients believed the earth to be flat and the center of the universe. The Greek Pythagoreans, for instance, taught that the earth was both round and rotating. The system's center was not the sun but a brilliant central fire that the sun reflected just as our moon (as we now know) "snatches" its "pale fire" (in Shakespeare's phrasing) from the sun. The earth, sun, moon, and the five other known planets circled the central fire. Since the earth always kept its uninhabited side toward the fire during its 24-hour revolution, the fire could never be seen. Aristotle suggested that it was the Pythagorean cult's obsession with the triangular number 10 (the sum of 1, 2, 3, 4) that led its members to add a 10th body called antichthon (counter-earth). It too was always invisible because its orbit lay between the earth and the central fire. Aristarchus of Samos, a third-century-B.C. Greek astronomer, actually proposed a heliocentric model, with all planets circling the sun, although his treatise on this was lost and is known only through comments by Archimedes.

The model that dominated Greek astronomy, however, as well as medieval science was the geocentric model of Aristotle: an unmoving spherical earth at the core of the universe with all other heavenly bodies, including the stars, going around it. Aristotle defended an earlier and excellent argument for the earth's roundness. During a lunar eclipse the earth's shadow on the moon has a rounded edge that can best be explained if the earth is a ball. The Ptolemaic model of the second century A.D., a refinement of Aristotle's, was designed to account for the erratic paths of the five visible planets as they cross our sky. The trick was done by having the planets move in smaller circles, called epicycles, as they travel larger circular orbits around the earth. The model was quite adequate to explain the apparent motions of heavenly bodies, including irregular movements of planets and moons caused by elliptical orbits, provided that enough epicycles were posited and bodies were allowed to move along them at nonuniform speeds.

We all know how, after a long controversy culminating in Galileo's persecution, the heliocentric model of the 16th-century Polish astronomer Nicolaus Copernicus finally won out. It is sometimes argued that it won only because it was simpler and more elegant. Thomas S. Kuhn has gone even further, and denied that the Copernican model was either simpler or more observationally accurate. ". . . The real appeal of sun-centered astronomy," he writes in *The Copernican Revolution*, was aesthetic rather than pragmatic. To astronomers the initial choice between Copernicus' system and Ptolemy's could only be a matter of taste. . . ."

But Kuhn is wrong. There were many astronomical observations, pointed out by Copernicus himself, for which his theory provided a much simpler explanation than did Ptolemy's, thus providing his theory with a superiority that was more than just "taste." Later, of course, it explained an enormous variety of astronomical phenomena, such as the bulging of the earth's equator, that the Ptolemaic theory could not account for. (On this see "Kuhn and the Copernican Revolution," by Richard J. Hall, in the *British Journal for the Philosophy of Science*, May 1970, pages 196–97.)

The final twist to this vacillating history came with Einstein's general theory of relativity. If this theory is correct, there are no absolute motions with respect to a fixed space and therefore no "preferred frame of reference." One can assume that the earth is fixed—not even rotating—and the tensor equations of relativity will account for everything. The earth is fat around its waist not because of inertial forces but because the rotating cosmos produces a gravitational field that causes the bulge. Since all motion is relative, the choice of a sun-centered model over an earth-centered one for the solar system is one of convenience. We say the earth rotates because it is enormously simpler to make the cosmos a fixed inertial frame of reference than to say it is rotating and shifting around in peculiar ways. It is not that the heliocentric theory is "truer." Indeed, the sun itself is moving and is in no sense the center of the cosmos, if indeed the cosmos has a center. The only "true" motion is the relative motion of the earth and the cosmos.

This arbitrariness about the frame of reference is involved in a funny argument that still pops up in parlor conversations. The moon circles the earth, as the earth circled the central fire in the Pythagorean model, so that it always keeps the same face toward the earth. This has intrigued poets, major and minor, as well as astronomers. Robert Browning's "One Word More" likens the moon's two sides to the two "soul-sides" of every man: "one to face the world with, one to show a woman when he loves her!" Edmund Gosse claimed that his housekeeper penned the following immortal quatrain:

> *O moon, when I gaze on thy beautiful face,*
> *Careering along through the boundaries of space,*
> *The thought has often come into my mind*
> *If I ever shall see thy glorious behind.*

The moon's habit of concealing its backside raises the following trivial question. Does the moon "rotate" as it goes around the earth? An astronomer would say yes, once for each revolution. It is hard to believe, but intelligent men have been so incensed by this assertion that they have published (usually at their own expense) lengthy pamphlets arguing that the moon does not rotate at all. (Several such treatises are discussed in Augustus De Morgan's *Budget of Paradoxes*.) Even the great Johannes Kepler preferred to think of the moon as nonrotating. He compared it to a ball fastened to a thong and whirled around the head. The sun rotates, he reasoned, to impart motion to its planets, and the earth rotates to impart motion to its moon. Since the moon has no smaller moon of its own, there is no need for it to rotate.

The problem of the moon's rotation is basically the same as a penny paradox described in Chapter 2 of my *Mathematical Carnival*. If you roll one penny around a fixed penny, keeping the rims together to prevent sliding, the rolling penny rotates twice during one round trip.

Or does it? Joseph Wisnovsky, an editor of *Scientific American*, has called my attention to a furious controversy over this question that raged in the letters department of this magazine for almost three years. In 1866 a reader asked: "How many revolutions on its own axis will a wheel make in rolling once around a fixed wheel of the same size?" "One," the editors replied. A torrent of correspondence followed from readers who disagreed. In Volume 18 (1868), pages 105–06, *Scientific American* printed a selection from "half a bushel" of letters supporting the double-rotation view. For the next three months the magazine published correspondence from both "oneists" and "dualists," including engravings of elaborate mechanical devices they had made and had sent to establish their case.

"If you swing a cat around your head," wrote oneist H. Bluffer (March 21, 1868), attacking the moon's rotation, "would his head, eyes and vertebrae each revolve on its own axis . . . ? Would he die at the ninth turn?"

The volume of mail reached such proportions that in April 1868 the editors announced they were dropping the topic but would continue it in a new monthly magazine, *The Wheel*, devoted to the "great question." At least one issue of this periodical appeared, because *Scientific American* readers were told in the May 23 issue that they could obtain *The Wheel* at newsstands or by mail for 25 cents. Perhaps the controversy was a put-on by the editors. Obviously it is no more than a debate over how one chooses to define the phrase "rotates on its own axis." To an observer on the fixed penny the moving coin rotates once. To an observer looking down from above it rotates twice. The moon does not rotate relative to the earth; it does rotate relative to the stars. Can the reader decide, without making a model, how many times the outside coin will rotate per one revolution (relative to you as the observer) if its diameter is half that of the fixed coin?

The same ridiculous question about lunar rotation could have been asked from 1890 to 1965 about Mercury. The Italian astronomer Giovanni Schiaparelli (the man who started all the nonsense about Martian irrigation canals by drawing maps of straight lines he imagined he saw crisscrossing the planet) announced in the late 1880's that his observations proved that Mercury always kept the same face toward the sun. In other words, it rotated once for every revolution of 88 days. For the next 75 years hundreds of observations by other eminent astronomers confirmed this. Because Mercury lacks an atmosphere to transfer heat it was assumed that its illuminated side was perpetually sizzling at 700 to 800 degrees Fahrenheit and that its dark side was perpetually close to absolute zero. "Mercury has the distinction," wrote Fred Hoyle as late as 1962, "of possessing not only the hottest place but also the coldest place in the whole planetary system."

Between Mercury's hot and cold sides there would of course be a girdle of everlasting gloaming, presumably with a climate mild enough to support life. The notion long intrigued writers of science fiction. "Twilight. Always twilight," says a visitor to Mercury in Arthur Jean Cox's 1951 story "The Twilight Planet." "The days pass, or so the clocks, the calendars tell you. But time, subjective time, is frozen delicately in midflight. The valley is an ocean of shadows; shade-tides lap upon the shores of mountains." In Robert Silverberg's "Sunrise on Mercury" (1957), astronauts land on Mercury's "Twilight Belt" between "the cold, ice-bound kingdom of Dante's deepest pit" and "the brimstone empire." The belt is a region where fire and frost meet, "each hemisphere its own kind of hell." When the story appeared in the 1969 Dell paperback anthology *First Step Outward*, editor Robert Hoskins had to append a note saying it had now passed from science fiction into the realm of fantasy.

The first hint that something was wrong came in 1964 when radio-telescope observations by Australian astronomers indicated that the supposed frozen side of Mercury has a temperature of about 60 degrees Fahrenheit! Could the planet, they wondered, have an atmosphere after all? In 1965 Gordon H. Pettengill and Rolf B. Dyce, using radar reflections from opposite edges of the planet, discovered the real reason. Schiaparelli had been as wrong about Mercury's spin as he had been about Mars's canals. Mercury rotates once every 59 days, exactly two-thirds of its orbital period. Apparently the little planet has a lopsided mass, like our moon, or a tidal bulge that allowed its capture by the sun in a stable 3/2 "resonance lock." For every two orbits it spins three times. One reason astronomers had been wrong for 75 years was that they usually looked at Mercury at a favorable time that occurs once a year. Because they always saw the same dusky markings, they assumed that, since Mercury had made four orbits, it had rotated four times when actually it had rotated six. Although such rationalizations can be made, wrote Irwin I. Shapiro [see "Radio Observations of the Planets," by Irwin I. Shapiro, in *Scientific American*, July 1968], it is still "unsettling to contemplate this persistence of self-deception." How Charles Fort, that eccentric iconoclast of science, would have gloated over such a gigantic goof!

Still more astonishing was the discovery made in 1962 about the spin of Venus. Its slow spin was believed to be so close to its orbital period of about 225 earth days that many astronomers (Schiaparelli for one) were convinced that, like Mercury and our moon, Venus had identical rotational and orbital periods. In 1962 astronomers using the Goldstone radar of California's Jet Propulsion Laboratory established two incredible facts. Venus spins slowly *backward* with respect to all other planets. (Uranus's direction of spin is ambiguous. Its axis is so close to being parallel to the plane of the ecliptic that either pole can be called north.) Venus is the only planet on which the sun rises (very slowly) in the west. Moreover, its spin period of 243.16 days—making its day longer than its year—is just such that, whenever Venus is closest to the earth, it always presents the same face toward *us!* In *Don Juan* Lord Byron speaks of "a rosy sky, with one star [Venus] sparkling through it like an eye." Why Venus should keep her eye on the earth in such a curious fashion is still a mystery. Presumably, like Mercury, it either is asymmetric in mass or has a large enough tidal bulge to have allowed capture by the earth in this unexpected resonance lock.

The story of Venus's nonmoon is another Keystone Cops episode in astronomical history. In 1645 an Italian astronomer, Francesco Fontana, asserted he had seen a moon of Venus. His observation was substantiated in 1672 by Jean Dominique Cassini, who had discovered two satellites of Saturn and was later to find two more. Venus's moon was also seen by many leading astronomers of the 18th century. The famous German mathematician, physicist, and astronomer Johann Heinrich Lambert published in 1773 a treatise on Venus's moon in which he even calculated its orbit. Frederick the Great honored Jean Le Rond d'Alembert by naming the moon after him, although the great French mathematician politely refused the honor. Of course there never was such a moon or it would have been visible as a black speck when Venus crossed the sun's disk. The astronomers had either seen nearby stars or ghost images produced by lens refraction or, as in the case of those astronomers who "saw" Martian canals, their hopes and beliefs were playing psychological tricks on their vision. Similar explanations surely account for many 18th- and 19th-century "observations" of Vulcan, a planet supposedly inside Mercury's orbit.

How did the solar system evolve? No one is sure. The most popular view at present is the one first advanced by Immanuel Kant. Somehow the planets condensed from gases and dust particles in a whirling disklike cloud that once surrounded the sun. The counterclockwise spin of this cloud, when viewed from above north poles, would explain why all the planets and most of their moons revolve in the same direction. Why, though, are the ancient tracks spaced the way they are? Is it mere happenstance or are their distance ratios governed by a mathematical law?

It was Kepler who dreamed up the most fantastic explanation. He first tried inscribing and circumscribing regular polygons, then spheres and cubes, but he failed to hit on a pattern that gave the right ratios. Suddenly an inspiration struck him. There are six planets, therefore five spaces between them. Are there not five and only five regular convex solids? By nesting the five Platonic solids inside one another in a certain order, with shells between them to take care of eccentricities in the planets' elliptical paths, he arrived at a structure that corresponded roughly with what were then believed to be the maximum and minimum distances of each planet from the sun [see Figure 85]. It

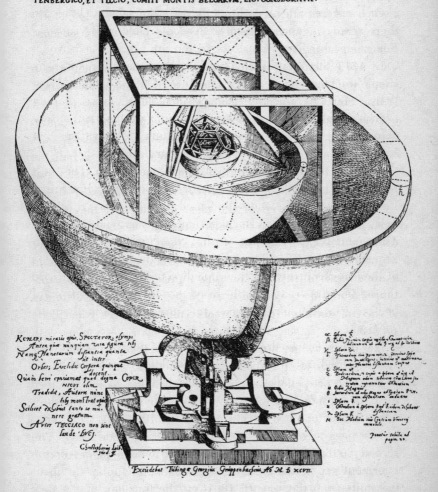

FIGURE 85

Kepler's model of the solar system

was a crazy theory, even in Kepler's time, but Kepler was a remarkable blend of stupendous scientific intuition and occult beliefs (including astrology) that led him to expect such geometric harmonies. "The intense pleasure I have received from this discovery," he wrote, "can never be told in words." Ironically, his correct convictions that the planets move in ellipses, not circles, and that tides are caused by the moon seemed so equally preposterous that even Galileo dismissed both views as more Keplerian fantasy.

In 1772 Johann Daniel Titius of Wittenberg announced a simple number sequence that seemed to fit the planetary orbits. It soon became known as "Bode's law" because four years later a more famous German astronomer, Johann Elert Bode, published the sequence in a textbook. To obtain the numbers, start with 0, 3, 6, 12, 24, 48, 96, 192, . . . Every number is half the next one except for 0, which really should be $1\frac{1}{2}$. To each number add 4. The resulting sequence—4, 7, 10, 16, 28, 52, 100, 196 . . . —gives the ratios of the mean distances of the planets from the sun. If we take the earth's distance as our "astronomical unit," the third number, 10, becomes 1. Dividing the other numbers by 10 then gives the mean distances of the planets in astronomical units. The chart in Figure 86 shows these distances alongside the actual ones. Note that the mean distances of the first six planets, still the only planets known when Bode published his paper, are in remarkably close agreement with values given by the Bode series. Not only that, but Bode's law succeeded in making two excellent predictions.

The first prediction was that a planet should be at a distance of 19.6 astronomical units. When Uranus was discovered in 1781, it was found to have a distance of 19.2, a fact that convinced most astronomers of the soundness of Bode's law. The second prediction was that there ought to be a planet in the enormous gap between the orbits of Mars and Jupiter, at about 2.8 units from the sun. In 1801, on the first day of the new century, Ceres, the largest of the asteroids, was discovered at

Planet	Bode's Series	Actual Mean Distance
MERCURY	.4	.39
VENUS	.7	.72
EARTH	1	1
MARS	1.6	1.52
(CERES)	2.8	2.77
JUPITER	5.2	5.20
SATURN	10	9.57
URANUS	19.6	19.15
NEPTUNE	38.8	29.95
PLUTO	77.2	39.39

FIGURE 86

Bode's law for the spacing of planetary orbits

2.77 units from the sun! Thousands of smaller planetoids were later observed in this region. Defenders of Bode's law argued plausibly that the planetoids were remnants of an exploded planet that had once orbited the sun at a spot close to where Bode's law said it should be.

Alas, the law failed for Neptune and Pluto, persuading many astronomers that the law's earlier successes had been accidental. Other astronomers have recently suggested that Pluto may be an escaped moon of Neptune and that before the two bodies separated Neptune could have been near the spot predicted by Bode's law. It has also been argued that Bode's law may apply to all planets except those at the inner and outer fringes of the solar system, where irregularities would be more likely. Since Mercury and Pluto have orbits much more eccentric and more inclined to the plane of the ecliptic than those of the other planets, it is not unreasonable to suppose fringe conditions would make them exceptions to a general rule.

Is Bode's law a numerological curiosity, as irrelevant as Kepler's nested polyhedrons, or does it say something of value that eventually will be explained by a theory of the solar system's origin? The question is still undecided. Defenders of the law usually cite the number sequence, announced in 1885 by the Swiss mathematician Johann Jakob Balmer, that fitted the frequencies of the spectrum lines of hydrogen. This series was pure numerology until decades later, when Niels Bohr found the explanation for "Balmer's series" in quantum mechanics.

"The question is," writes Irving John Good in a recent paper on Bode's law that is listed in this book's bibliography, "whether a piece of scientific numerology unsupported by a model is sufficiently striking to make us say that people ought to look for a scientific model in order to explain it." From my layman's seat I hesitate even to guess how Bode's law will fare in future years.

I conclude with another tricky problem. As the earth goes around the sun, its moon traces a wavy path with respect to the sun. How many sections of that wavy path, during 12 lunar orbits around the earth, are concave in the sense that their convex sides are toward the sun?

ANSWERS

THE ANSWER to the first problem is that a wheel rotates three times in rolling once around a fixed wheel with twice the diameter of the rolling wheel. Since the circumference of the rolling wheel is half that of the larger one, this produces two rotations with respect to the fixed wheel, and the revolution adds a third rotation with respect to an observer from above. The general formula, where a is the diameter of the fixed wheel and b is the diameter of the rolling wheel, is $(a/b) + 1$. This gives the number of rotations for one revolution. Thus if the rolling wheel has a diameter twice that of the fixed wheel, it rotates $1\frac{1}{2}$ times. The rolling wheel, as it gets larger, approaches a limit of one rotation per revolution—a limit that is achieved only when it rolls around a degenerate "circle" of zero diameter, namely a point. Suppose the diameter of the fixed coin equals the circumference of the revolving coin. How many times does the outside coin rotate per revolution?

The answer to the question about the number of concave sections in the moon's wavy path around the sun is that at *no* time is the path concave. The moon is so close to the earth and the earth's speed is so great in relation to the moon's speed around the earth that the moon's path (in relation to the sun) is at all points convex.

CHAPTER 17

Mascheroni Constructions

IT IS OFTEN SAID that the ancient Greek geometers, following a tradition allegedly started by Plato, constructed all plane figures with a compass and a straightedge (an unmarked ruler). This is not true. The Greeks used many other geometric instruments, including devices that trisected angles. They did believe, however, that compass-and-straightedge constructions were more elegant than those done with other instruments. Their persistent efforts to find compass-and-straightedge ways to trisect the angle, square the circle, and duplicate the cube—the three great geometric construction problems of antiquity—were not proved futile for almost 2,000 years.

In later centuries geometers amused themselves by imposing even more severe restrictions on instruments used in construction problems. The first systematic effort of this kind is a work ascribed to the 10th-century Persian mathematician Abul Wefa, in which he described constructions possible with the straightedge and a "fixed compass," later dubbed the "rusty compass." This is a compass that never alters its radius. The familiar

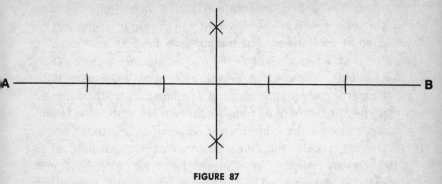

FIGURE 87
How to bisect a line of any length with a "rusty compass"

methods of bisecting a line segment or an angle are simple examples of fixed-compass-and-straightedge constructions. Figure 87 shows how easily a rusty compass can be used for bisecting a line more than twice the length of the compass opening. Many of Abul Wefa's solutions—in particular, his method of constructing a regular pentagon, given one of its sides—are extremely ingenious and hard to improve on.

Figure 88 shows how a rusty compass can be used for drawing a line parallel to line *AB* and through any point *P* outside

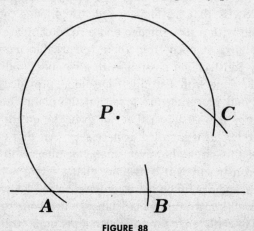

FIGURE 88
A rusty-compass construction of a parallel line

the line. This is done by constructing the corners of a rhombus in three steps, and is so simple that you can figure it out just by looking at the picture. The method goes back at least to 1574, yet it is still being rediscovered and written up as new. (See, for example, *Mathematics Teacher*, February 1973, page 172.)

Leonardo da Vinci and numerous Renaissance mathematicians experimented with fixed-compass geometry, but the next important treatise on the subject was *Compendium Euclidis Curiosi*, a 24-page booklet published anonymously in Amsterdam in 1673. It was translated into English four years later by Joseph Moxon, England's royal hydrographer. This work is now known to have been written by a Danish geometer, Georg Mohr, whom we shall meet again in a moment. In 1694 a London surveyor, William Leybourn, in a whimsical book called *Pleasure with Profit*, treated rusty-compass constructions as a form of mathematical play. Above his section on this topic he wrote: "Shewing How (Without Compasses), having only a common Meat-Fork (or such like Instrument, which will neither open wider, nor shut closer), and a Plain Ruler, to perform many pleasant and delightful Geometrical Operations."

In the 19th century the French mathematician Jean Victor Poncelet suggested a proof, later rigorously developed by Jacob Steiner, a Swiss, that *all* compass-and-straightedge constructions are possible with a straightedge and a fixed compass. This conclusion follows at once from their remarkable demonstration that every construction possible with a compass and a straightedge can be done with a straightedge alone, provided that a single circle and its center are given on the plane. Early in the 20th century it was shown that not even the entire "Poncelet-Steiner circle," as it is called, is necessary. All that is needed is one arc of this circle, however small, together with its center! (In such constructions it is assumed that a circle is constructed if its center and a point on its circumference are determined.)

Many well-known mathematicians studied constructions that are possible with such single instruments as a straightedge, a straightedge marked with two points, a ruler with two parallel straightedges, a "ruler" with straightedges meeting perpendicularly or at other angles, and so on. Then in 1797 an Italian geometer, Lorenzo Mascheroni, amazed the mathematical world by publishing *Geometria del compasso*, in which he proved that every compass-and-straightedge construction can be done with a movable compass alone. Since straight lines cannot of course be drawn with a compass alone, it is assumed that two points, obtained by arc intersections, define a straight line.

Compass-only constructions are still called Mascheroni constructions even though it was discovered in 1928 that Mohr had proved the same thing in an obscure little work, *Euclides Danicus*, published in 1672 in Danish and Dutch editions. A Danish student who had found the book in a secondhand bookstore in Copenhagen showed it to his mathematics teacher, Johannes Hjelmslev of the University of Copenhagen, who instantly recognized its importance. Hjelmslev published it in facsimile, with a German translation, in Copenhagen in 1928.

Today's geometers have little interest in Mohr-Mascheroni constructions, but because they present so many problems of a recreational nature, they have been taken over by puzzle enthusiasts. The challenge is to improve on earlier constructions by finding ways of doing them in fewer steps. Sometimes it is possible to improve on Mohr's or Mascheroni's methods, sometimes not. Consider, for example, the simplest of five solutions by Mascheroni to his problem No. 66, that of finding a point midway between two given points A and B [*see Figure 89*].

Draw two circles of radius AB, their centers at A and B. Keeping the same compass opening, with C and D as centers, mark points D and E. (Readers may recall that this is the beginning of the well-known procedure by which a circle is divided into six equal arcs, or three equal arcs if alternate points are taken.) Point E will lie on an extension of line AB to the right, and AE will be twice AB. (This procedure obviously can be repeated rightward to double, triple, or produce any multiple

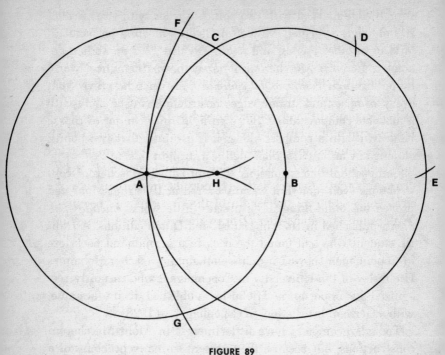

FIGURE 89

*Mascheroni's method of finding point H, midway between
A and B, using a compass alone*

of length *AB*.) Open the compass to radius *AE* and draw an
arc, its center at *E*, that intersects the left circle at *F* and *G*.
Close the compass to radius *AB* once more. With centers at *F*
and *G*, draw the two arcs that intersect at *H*.

H is midway between *A* and *B*. This is easily proved by
noting that the two isosceles triangles marked by corners *AFH*
and *AFE* share the base angle *FAE* and therefore are similar.
AF is half of *AE*; consequently *AH* is half of *AB*. For readers
acquainted with inversion geometry, *H* is the inverse of *E* with
respect to the left circle. A simple proof of the construction, by
way of inversion geometry, is in *What Is Mathematics?*, by
Richard Courant and Herbert Robbins (Oxford, 1941), page
145. Note that if line segment *AB* is drawn at the outset, and

the problem is to find its midpoint with compass alone, only one of the last two arcs need be drawn, reducing the number of steps to six. I know of no way to do this with fewer steps.

Another famous problem solved by Mascheroni is locating the center of a given circle. His method is too complicated to reproduce here, but fortunately a simplified approach, of unknown origin, appears in many old books and is given in Figure 90. *A* is any point on the circle's circumference. With *A* as the center, open the compass to a radius that will draw an arc intersecting the circle at *B* and *C*. With radius *AB* and centers *B* and *C*, draw arcs that intersect at *D*. (*D* may be inside or below the circle, depending on the length of the first compass opening.) With radius *AD* and center *D*, draw the arc giving intersections

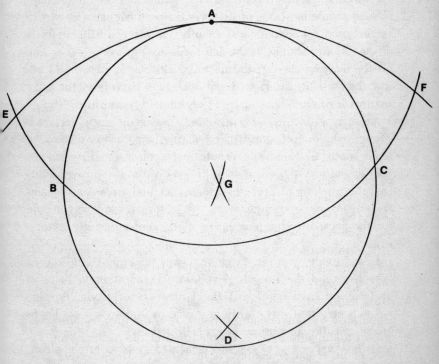

FIGURE 90

How to find a circle's center, using the compass alone, in six steps

E and *F*. With radius *AE* and centers *E* and *F*, draw arcs intersecting at *G*. *G* is the circle's center. As before, there is an easy proof that starts by observing that the two isosceles triangles marked by *DEA* and *GEA* share the base angle *EAG* and are therefore similar. For the rest of the proof, as well as a proof by inversion geometry, see L. A. Graham's *The Surprise Attack in Mathematical Problems* (Dover, 1968), problem No. 34.

A third well-known problem in Mascheroni's book has become known as "Napoleon's problem" because it is said that Napoleon Bonaparte originally proposed it to Mascheroni. It is not generally known that Napoleon was an enthusiastic amateur mathematician, of no great insight but particularly fascinated by geometry, which of course had great military value. He was also a man with unbounded admiration for the creative French mathematicians of his day. Gaspard Monge (known to recreational mathematicians mainly for his youthful analysis of the "Monge shuffle," in which cards are pushed one at a time off the deck by the left thumb to go alternately above and below the cards in the right hand) seems to have been the only man with whom Napoleon had a permanent friendship. "Monge loved me as one loves a mistress," Napoleon once declared. Monge was one of several French mathematicians who were made counts by Napoleon. Whatever Napoleon's ability as a geometer may have been, it is to his credit that he so revolutionized the teaching of French mathematics that, according to several historians of mathematics, his reforms were responsible for the great upsurge of creative mathematics in 19th-century France.

Like Monge, young Mascheroni was an ardent admirer of Napoleon and the French Revolution. In addition to being a professor of mathematics at the University of Pavia, he also wrote poetry that was highly regarded by Italian critics. There are several Italian editions of his collected verse. His book *Problems for Surveyors* (1793) was dedicated in verse to Napoleon. The two men met and became friends in 1796, when Napoleon invaded northern Italy. A year later, when Mascheroni published his book on constructions with the compass alone, he again honored Napoleon with a dedication, this time a lengthy ode.

Napoleon mastered many of Mascheroni's compass constructions. It is said that in 1797, while Napoleon was discussing geometry with Joseph Louis Lagrange and Pierre Simon de Laplace (famous mathematicians whom Napoleon later made a count and a marquis respectively), the little general surprised them by explaining some of Mascheroni's solutions that were completely new to them. "General," Laplace reportedly remarked, "we expected everything of you except lessons in geometry." Whether this anecdote is true or not, Napoleon did introduce Mascheroni's compass work to French mathematicians. A translation of *Geometria del compasso* was published in Paris in 1798, a year after the first Italian edition.

"Napoleon's problem" is to divide a circle, its center given, into four equal arcs, using the compass alone. In other words, find the four corners of an inscribed square. A beautiful six-arc solution is shown in Figure 91. With the compass open to the circle's radius, choose any point A, then mark spots B, C, and D, using A, B, and C as centers. Open the compass to radius AC. With centers A and D, draw the arcs intersecting at E. With center A and radius OE, draw the arc that cuts the original circle at F and G. A, F, D, and G are the corners of the inscribed square. I do not know if this is Mascheroni's solution (his book has not been translated into English and I have not had access to the Italian or French editions) or a later discovery. Henry Ernest Dudeney gives it without proof in *Modern Puzzles* (1926). A simple proof can be found in Charles W. Trigg's *Mathematical Quickies* (McGraw-Hill, 1967), problem No. 248.

Two related and less well-known Mascheroni problems are: (1) Given two adjacent corner points of a square, find the other two, and (2) Given two diagonally opposite corner points of a square, find the other two. An eight-arc solution to the first problem was separately sent to me by readers Don G. Olmstead

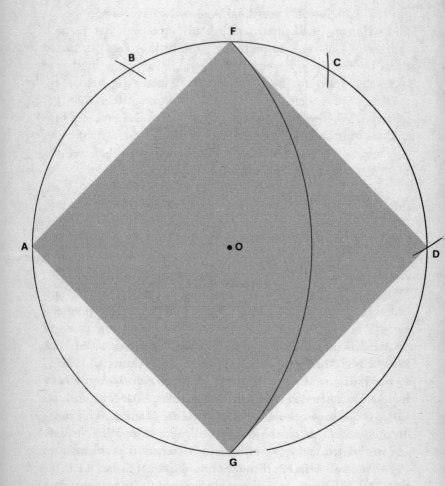

FIGURE 91
Six-step solution to "Napoleon's problem"

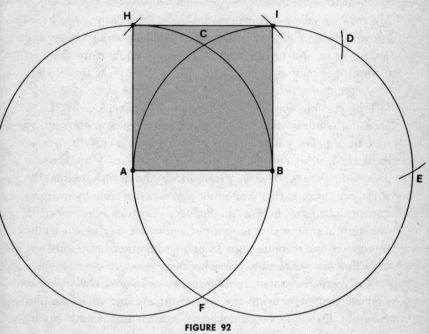

FIGURE 92

An eight-step way to construct the corners of a square,
given adjacent corners A and B

and Paul White and can be found, with a proof, in M. H. Greenblatt's *Mathematical Entertainments* (Crowell, 1965), page 139. Figure 92 shows the procedure. *A* and *B* are the two given points. After drawing the two circles, each with radius *AB*, keep the same opening and mark points *D* and *E*, with centers at *C* and *D*. Open the compass to radius *CF*. With *A* and *E* as centers, draw the two arcs that intersect at *G*. With radius *GB* and centers *A* and *B*, draw the arcs that cut the circles at *H* and *I*. *H* and *I* are the two corner points sought.

The best solution I know for the second and more difficult problem requires nine arcs. Readers are invited to search for it, or a better one.

ADDENDUM

MANNIS CHAROSH called my attention to the surprising, little-known theorem that all points obtainable by straightedge and compass can also be obtained by using nothing more than an unlimited supply of identical toothpicks. The picks model rigid line segments which can be moved about on the plane.

This curious construction method was invented by T. R. Dawson, editor of the *Fairy Chess Review*, and written up by him in a paper called " 'Match-Stick' Geometry," in *Mathematical Gazette*, Volume 23, May 1939, pages 161–68. Dawson proves the general theorem given above, and also shows that the sticks cannot construct points not also constructable by compass and straightedge. He gives methods for bisecting a line segment, bisecting an angle, dropping a perpendicular, laying a parallel to a given line through a given point, and other basic constructions that are sufficient to prove his case.

The recreation raises innumerable unexplored challenges to find constructions using the minimum number of sticks. For example, Dawson's best method of constructing a unit square (one with a side equal to the length of a stick) is shown in Figure 93, where AF is any line within the angle BAC. It uses 16 sticks.

Dawson asserts that 11 sticks are minimal for bisecting a given line of unit length, and 13 sticks for finding the midpoint between two given points a unit distance apart. He challenges the reader to find a 10-stick method of determining the midpoint between two points that are more distant than the unit length of a stick, but less distant than the square root of 3.

Figure 94 shows a simple 5-stick method of bisecting any angle not greater than 120 degrees and not exactly 60 degrees. The method also constructs a perpendicular from C to line AB. Lines are extended, and parallel lines constructed, simply by extending a row of side-by-side equilateral triangles as far as desired.

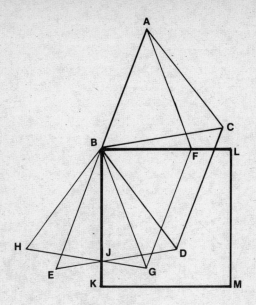

FIGURE 93
Constructing a square with 16 toothpicks

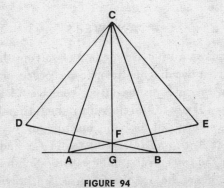

FIGURE 94
Bisecting an angle with five toothpicks

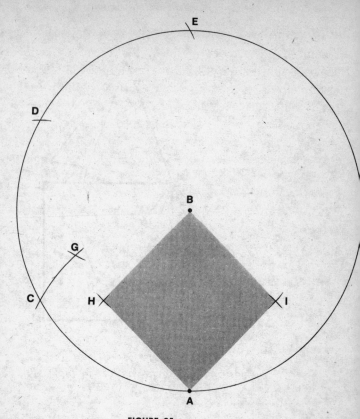

FIGURE 95

Construction of a square, given diagonal corners A *and* B

ANSWERS

FIGURE 95 shows a nine-arc method of solving the Mascheroni problem: Given two diagonally opposite corners of a square, find the other two corners using only a compass. *A* and *B* are the given corners. Draw the circle with radius *AB* and *B* as center. Keeping the compass at the same opening, draw arcs *C*, *D*, and *E* (centers at *A*, *C*, and *D*). With radius *CE* and centers at *A* and *E*, draw the two arcs that intersect at *F*. With radius *BF* and center at *E*, draw the arc that intersects a previous arc at *G*.

With radius *BG* and centers at *A* and *B*, draw the arcs intersecting at *H* and *I*. *A*, *H*, *B*, and *I* are the corners of the desired square. Philip G. Smith, Jr., of Hastings-on-Hudson, N.Y., sent a simple proof of the construction, based on right triangles and the Pythagorean theorem, but I leave it to interested readers to work out such proofs for themselves.

After writing my column on Mascheroni constructions I learned that the six-arc solution to "Napoleon's problem" is indeed Mascheroni's. Fitch Cheney sent me his paper "Can We Outdo Mascheroni?" (*The Mathematics Teacher*, Vol. 46, March 1953, pages 152–56), in which he gives Mascheroni's solution followed by his own simpler solution using only five arcs.

Cheney's solution is shown in Figure 96. Pick any point *A* on the given circle and draw a second circle with radius *AO*. With *C* as center and the same radius, draw a third circle. With *D* as center and radius *DA*, draw the arc intersecting the original circle at *E*. With *F* as center and radius *FO*, draw an arc crossing the preceding arc at *G*. With *C* as center and radius *CG*, draw the arc intersecting the original circle at *H* and *I*. *E*, *I*, *C*, and *H* mark the corners of the desired square.

Cheney calls attention in his article to the difference between a "modern compass," which retains its opening like a divider, and the "classical compass" of Euclid, which closes as soon as either leg is removed from the plane. Cheney's five-arc solution uses only classical arcs, in contrast to Mascheroni's. Cheney also gives in his article a seven-step classical method of inscribing a pentagon in a circle, two steps fewer than Mascheroni's modern-compass method.

A large number of readers noticed that Mascheroni's compass-only method of constructing a point midway between two points can be reduced by one step. The distance between the intersections of the two circles of Figure 89 clearly is equal to *CE*, therefore point *E* can be found without the intermediate step of finding point *D*. This procedure, as many readers pointed out, automatically lowers by one the number of arcs required

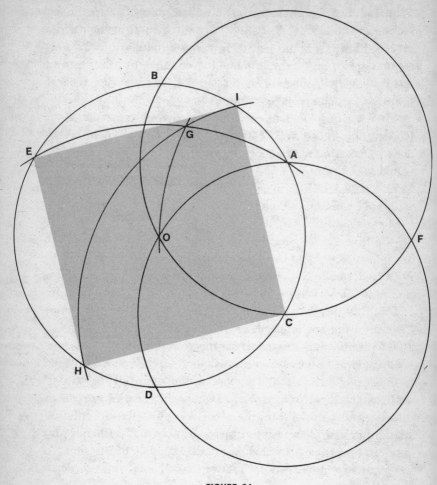

FIGURE 96

Fitch Cheney's simpler solution for "Napoleon's problem"

to bisect a line segment, to find the four corners of a square inscribed in a given circle ("Napoleon's problem"), and, given adjacent corners of a square, to find the other two corners.

The problem of finding the other two corners of a square

when given two diagonally opposite corners—a problem I answered with nine arcs—reduces to eight arcs by adopting the procedure just described. However, more than a dozen readers discovered a beautiful six-arc solution [*see Figure 97*]. *A* and *B* are the given corners. After drawing the two circles through these two points, open the compass to *CD*, and with *C* as center draw arc *EDF*. With *F* as center and *AF* as radius, draw arc *GAH*. With *E* and *F* as centers and *EG* as radius, draw the two arcs intersecting at *X* and *Y*. It is not hard to prove that *AXBY* are the corners of the desired square.

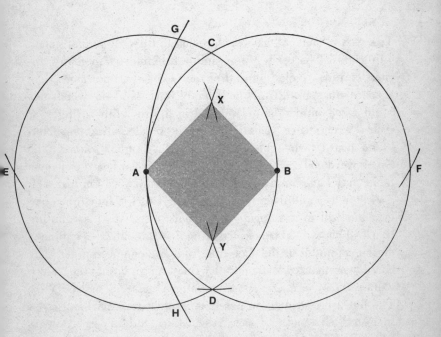

FIGURE 97
A six-arc solution to a Mascheroni problem

CHAPTER 18

The Abacus

THE WORD "ABACUS" has been applied to three dissimilar calculating aids. The earliest and simplest, employed in many ancient cultures, including that of the Greeks, was no more than a board dusted with a thin layer of dark sand on which one could trace numerals and geometric figures with a finger or stylus. Archimedes was said to have been calculating on such a "sand board" when he was killed by a Roman soldier. The Greek word *abax*, which meant in general a flat board or legless table, may have come from *abaq*, the Hebrew word for dust.

A later type of abacus, known as early as the fourth century B.C. and still in use during the Renaissance, was the counting board. This was a true calculating instrument, as genuine a digital computer as the slide rule is an analogue computer. The board was marked with parallel lines representing the "place values" of a number system, usually a notation based on 10. The lines were drawn on parchment, etched on marble, carved in wood, or sometimes even stitched in cloth. Loose counters were moved back and forth on these lines to perform simple calculations. The Greeks called the board an *abakion*; the Romans called it an *abacus*. The counters were round pebbles or similar objects that were moved along grooves, and the Latin word for pebble, *calculus*, is therefore the origin of such words as "calculate" and "calculus." Several pictures, one on a Greek vase, show the counting board in use, but only one Greek counting board survives: a marble rectangle measuring about five by six inches that was found on the island of Salamis. During the Middle Ages checkered counting boards were in general use, and that explains the origin of such words as "check" and "exchequer."

The device we now know as an abacus is essentially a counting board modified so that the counters are set in grooves or slide along wires or rods. It is of unknown origin. The ancient Greeks probably did not have such instruments; the earliest references to them are in Roman literature. The counters, which the Romans called *claviculi* (little nails), moved up and down in grooves. The Romans had several forms of the device. A small bronze abacus used in Italy as late as the 17th century is of particular interest because its basic structure is the same as today's Japanese abacus. Each vertical groove stands for a power of 10, the powers increasing serially to the left. Four counters in each groove below a horizontal bar represent unit multiples of the place value. One counter in each groove above the bar represents five times the place value.

Here we encounter a curious state of affairs stressed by Karl Menninger, the German mathematician, in his beautiful and encyclopedic work *Number Words and Number Symbols*. For more than 15 centuries the Greeks and Romans and then Europeans of the Middle Ages and early Renaissance calculated on devices with authentic place-value systems in which zero was represented by an empty line or groove or by an empty position on the line or groove. Yet when these same people calculated without mechanical aids, they used clumsy notational systems lacking both place values and zero. It took a long time, as Menninger says, to realize that in writing numbers efficiently it is necessary to draw a symbol to indicate that a place in the number symbolizes nothing.

Perhaps the main reason for this cultural mental block was that papyrus and parchment were hard to come by. Because calculating was done almost entirely on abaci, there was no pressing need for a better written notation. It was the Italian Leonardo of Pisa, known as Fibonacci, who introduced the Hindu-Arabic notation to Europe in 1202 (see page 152). This led to an acrimonious struggle between the "abacists," who clung to Roman numerals in written computation but calculated on abaci, and the "algorists," who discarded Roman numerals altogether for the superior Hindu-Arabic notation. "Algorist" derives from the name of a ninth-century Arabic writer on mathematics, al-Khowârizmî, and is the ancestor of the modern word "algorithm." (In Figure 98 an abacist is shown competing against an algorist. The print is from a 16th-century book, *Margarita Philosophica*.) In some European countries calculating by "algorism" actually was forbidden by law, so that it had to be done in secret. There was opposition to it even in some Arabic countries. Not until paper became plentiful in the 16th century did the new notation finally win out, and soon after that the shapes of the 10 digits became standardized because of printing.

The abacus was discarded gradually in Europe and England. Remnants of it survive in the U.S. today only as colored beads on playpens, as devices for teaching decimal notation in the early grades, and in such counting aids as the rosary and the overhead sliding beads for recording billiard scores. In a way this is a pity because in recent centuries calculating with the abacus has been developed into an art in Eastern countries and in Russia. It is a multisensory experience: the abacist sees the beads move, hears them click, and feels them, all at once. Surely no digital computer has such high reliability in proportion to such low cost of purchase and maintenance.

Three types of abacus are in constant use today. The Chinese

FIGURE 98
An "abacist" (right) competes against an "algorist"
in a 16th-century print.

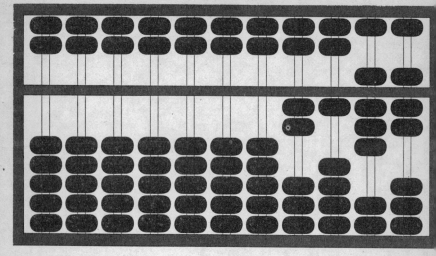

FIGURE 99

Chinese suan pan shows the number 2,187.

suan pan [*see Figure 99*], also used in Korea, has beads shaped like tiny doughnuts that move almost frictionlessly along bamboo rods. Each rod has five beads (ones) below the bar and two (fives) above. The Chinese symbol for *suan*, "calculate," is re-

produced from Menninger's book; it shows an abacus held below by the symbol for "hands," and with the symbol for "bamboo" above the abacus. The suan pan's origin is unknown. Precise descriptions go back to the 16th century but it is surely centuries older.

The Japanese soroban [*see Figure 100*] also can be traced to the 16th century, when it was probably borrowed from China.

Its counters are sharp-edged: two cones joined at their bases. Each rod has only one bead above the bar in a region the Japanese call "heaven," and only four below in the "earth." (The device originally had five beads below, like its Chinese counterpart, but the fifth was dropped about 1920. The two extra beads on each rod of the suan pan are not essential for modern abacus calculating, and discarding them produced a simpler instrument.) Japan still has yearly abacus contests in which thousands participate, and the soroban is still used in shops and small businesses, although it is rapidly being replaced in banks and large firms by modern desk computers.

There have been many contests in which a Japanese or Chinese abacist was pitted against an American operator of a desk computer. The most publicized was in 1946 in Tokyo, when Private Thomas Wood matched skills with Kiyoshi Matsuzaki. The abacist was faster in all calculations except the multiplication of huge numbers. One reason for the great speed of Oriental abacists, it should be admitted, is that they do a lot of work in their head, using the abacus mainly to record stages of the process.

The principal defect of abacus computation is that it preserves no record of past stages. If a mistake is made, the entire calculation has to be repeated. Japanese firms often ensure against this by having three abacists do the same problem simultaneously.

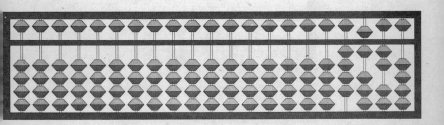

FIGURE 100
Modern Japanese soroban with the number 4,620 displayed

If all answers agree, it is assumed, following the rule given by the Bellman in Lewis Carroll's *The Hunting of the Snark*, that "What I tell you three times is true."

Russia's *s'choty* [*see Figure 101*] is markedly different from Oriental abaci. The Russians probably acquired it from the Arabs, and it is still used in parts of India and in the Middle East, where Turks call it a *coulba* and Armenians a *choreb*. In modern Russia the situation is the same as it is in Japan: almost every shopkeeper still uses an abacus, although it is being replaced in the accounting departments of large firms by modern desk calculators. The *s'choty* has horizontal wires or rods, most of them holding 10 beads; the two middle beads are of a differ-

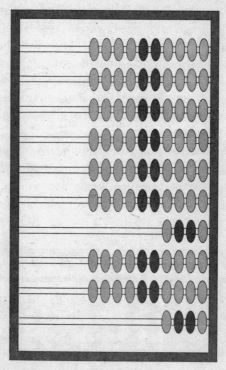

FIGURE 101
A Russian s'choty

ent color to make it easy to see where to divide them. The four-bead rods on the one shown in the illustration are used for fractions of rubles and kopeks.

In recent years the obvious value of the abacus for teaching arithmetic to blind children has been recognized and special abaci have been developed to reduce friction. Terrance V. Cranmer devised a soroban with foam rubber and felt under spherical beads that is available from The American Printing House for the Blind, 1839 Frankfort Avenue, Louisville, Ky. 40206. The firm also sells a manual in braille by Fred Gissoni. Victor E. Haas uses gravity to keep the beads of a soroban from accidentally sliding together by putting them on wire loops that curve upward in semicircles, a principle used in a less extreme way in some Russian abaci.

The easiest kind of calculation to master on the abacus is addition. For readers who lack the time or interest to learn the finger movements for subtraction (abacus movements must be automatic reflexes; it will not do to stop and think how to make addition movements in reverse), there is an old method of subtracting on the abacus by adding. Instead of subtracting the smaller number you add the "complement" of each of the digits with respect to 9. For example, you wish to take 9,213 from 456,789 on an Oriental abacus. Place 456,789 on the abacus. Mentally put two zeros in front of 9,213 to make it the same length as the other number. Then add pairs of digits in the usual manner, but from left to right (not the other way, as on paper), except that for each digit in 009,213 you substitute its difference from 9. In brief, to 456,789 you add 990,786. The result, 1,447,575, must now be given a final adjustment. Remove the single bead on the left and add one bead to the end digit on the right. This gives 447,576, the correct answer. In actual practice this final adjustment is sidestepped by not pushing up the bead on the left when you make the first addition and by raising an extra bead on the final addition. Supplying extra zeros at the left of a short subtrahend can also be avoided by remembering to remove a bead not from the extreme left but from the first digit to the left of the number of digits in the subtrahend.

Subtracting by adding complements is the method used in Comptometers as well as in high-speed electronic computers. The method applies to any number system, provided, of course, that the complements are with respect to the system's base less one. Thus for a 12-base system you add complements with respect to 11. For a computer using the binary system, complementation is simple because it is the same as changing every 1 to 0 and every 0 to 1. It goes without saying that abaci can be constructed for any base notation. The Oriental abaci adapt easily to certain other bases. For the binary, use only the heaven portion of the soroban. Its earth region can be used for the 5-base system. The suan pan can be similarly used for systems based on 3 or 6. For a 4-base system confine your attention to the top three beads below the bar of either device. For a 12-base system use the Chinese abacus, assigning a value of 6 instead of 5 to the beads above the bar.

An excellent practice exercise for abacus addition is linked to an old number stunt sometimes used by grade school teachers. The "magic number" 12,345,679 (note the missing 8) is chalked on the blackboard. A child is asked to step forward and name any digit. Suppose he picks 7. The teacher writes 63 below 12,345,679 and then asks the child to do the multiplication. It turns out, one hopes to everyone's amusement, that the product consists entirely of 7's. (To determine the multiplier the teacher simply multiplies the chosen digit by 9.)

To use this magic number as an abacus exercise, put 12,345,-679 on the abacus and add the same number to it eight times, the equivalent of multiplying it by 9. If your eight additions are done correctly, the abacus will show a row of 1's (single beads against the bottom of the bar). Add the magic number nine more times to produce a row of 2's. Nine more additions form a row of 3's, and so on until you finish, after 80 additions, with a row of 9's. Every finger movement is called into play by this exercise. Moreover, you can easily check the accuracy of your work at nine stages along the way, and by timing each stage you can tell how your speed improves from day to day.

An infinity of other magic numbers have the same property as 12,345,679 when multiplied by a product of any digit, d, and a certain constant. The product of 37 and $3d$ consists entirely of d's—for example, $37 \times (3 \times 8) = 888$. For $7d$ the smallest magic number is 15,873, for $13d$ it is 8,547, and for $99d$ it is 1,122,334,455,667,789. It is not hard to find such numbers. As an easy question, what is the smallest magic number for $17d$? In other words, what number when multiplied by $17d$, where d is any digit, gives a number consisting entirely of d's?

ANSWERS

THE PROBLEM was to find the smallest number that, when it is multiplied by $17d$, where d is any digit, gives a product that consists entirely of d's.

Such a number obviously must produce a string of 1's when multiplied by 17, therefore we divide 1111 . . . by 17 to see if we reach a point where there is no remainder. Such a point is first reached by the quotient 65,359,477,124,183, the answer to the problem. Since 17 times this number is 1,111,111,111,111,-111, a multiplier of $17 \times 2 = 34$ will produce a row of 2's, and so on for the remaining digits.

Because the infinite repeating decimal .1111 . . . equals 1/9, or the reciprocal of 9, it can be shown that each integral magic number is the repeating portion of the decimal form of a reciprocal of uneven multiples of 9 that are not multiples of 5. In this case the magic number is the repetend in the decimal of 1/153, the reciprocal of the product of 9 and 17. An example of a nonintegral magic number is 1.375. Multiplied by $8d$, the product consists entirely of d's, provided that the zeros at the right of the product's decimal point are disregarded.

CHAPTER 19

Palindromes:
Words and Numbers

A man, a plan, a canal—Suez!
—ETHEL MERPERSON,
a "near miss" palindromist,
in *Son of Giant Sea Tortoise*,
edited by Mary Ann Madden (Viking, 1975)

A PALINDROME is usually defined as a word, sentence, or set of sentences that spell the same backward as forward. The term is also applied to integers that are unchanged when they are reversed. Both types of palindrome have long interested those who amuse themselves with number and word play, perhaps because of a deep, half-unconscious aesthetic pleasure in the kind of symmetry palindromes possess. Palindromes have their analogues in other fields: melodies that are the same backward, paintings and designs with mirror-reflection symmetry, the bilateral symmetry of animals and man [*see Figure 102*] and so on. In this chapter we shall restrict our attention to number and language palindromes and consider some entertaining new developments in both fields.

An old palindrome conjecture of unknown origin (there are references to it in publications of the 1930's) is as follows. Start with any positive integer. Reverse it and add the two numbers.

FIGURE 102

Flying seagull: a visual palindrome

This procedure is repeated with the sum to obtain a second sum, and the process continues until a palindromic sum is obtained. The conjecture is that a palindrome always results after a finite number of additions. For example, 68 generates a palindrome in three steps:

$$
\begin{array}{r}
68 \\
+\ 86 \\
\hline
154 \\
+451 \\
\hline
605 \\
+506 \\
\hline
1,111
\end{array}
$$

For all two-digit numbers it is obvious that if the sum of their digits is less than 10, the first step gives a two-digit palindrome. If their digits add to 10, 11, 12, 13, 14, 15, 16, or 18, a palindrome results after 2, 1, 2, 2, 3, 4, 6, 6 steps respectively. As Angela Dunn points out in *Mathematical Bafflers* (McGraw-Hill, 1964), the exceptions are numbers whose two digits add to 17. Only 89 (or its reversal, 98) meets this proviso. Starting with either number does not produce a palindrome until the 24th operation results in 8,813,200,023,188.

The conjecture was widely regarded as being true until recently, although no one had succeeded in proving it. Charles W. Trigg, a California mathematician well known for his work on recreational problems, examined the conjecture more carefully in his 1967 article "Palindromes by Addition." He found

249 integers smaller than 10,000 that failed to generate a palindrome after 100 steps. The smallest such number, 196, was carried to 237,310 steps in 1975 by Harry J. Saal, at the Israel Scientific Center. No palindromic sum appeared. Trigg believes the conjecture to be false. (The number 196 is the square of 14, but this is probably an irrelevant fact.) Aside from the 249 exceptions, all integers less than 10,000, except 89 and its reversal, produce a palindrome in fewer than 24 steps. The largest palindrome, 16,668,488,486,661, is generated by 6,999 (or its reversal) and 7,998 (or its reversal) in 20 steps.

The conjecture has not been established for any number system, and has been proved false only in number notations with bases that are powers of 2. (See the paper by Heiko Harborth listed in the bibliography.) The smallest binary counterexample is 10110 (or 22 in the decimal system). After four steps the sum is 10110100, after eight steps it is 1011101000, after 12 steps it is 101111010000. Every fourth step increases by one digit each of the two sequences of underlined digits. Brother Alfred Brousseau, in "Palindromes by Addition in Base Two," proves that this asymmetric pattern repeats indefinitely. He also found other repeating asymmetric patterns for larger binary numbers.

There is a small but growing literature on the properties of palindromic prime numbers and conjectures about them. Apparently there are infinitely many such primes, although so far as I know this has not been proved. It is not hard to show, however, that a palindromic prime, with the exception of 11, must have an odd number of digits. Can the reader do this before reading the simple proof in the answer section? Norman T. Gridgeman has conjectured that there is an infinity of prime pairs of the form 30,103–30,203 and 9,931,399–9,932,399 in which all digits are alike except the middle digits, which differ by one. But Gridgeman's guess is far from proved.

Gustavus J. Simmons has written two papers on palindromic powers. After showing that the probability of a randomly selected integer being palindromic approaches zero as the number of digits in the integer increases, Simmons examines square numbers and finds them much richer than randomly chosen integers in palindromes. There are infinitely many palindromic squares, most of which, it seems, have square roots that also are palindromes. (The smallest nonpalindromic root is 26). Cubes too are unusually rich in palindromes. A computer check on all cubes less than 2.8×10^{14} turned up a truly astonishing fact. The only palindromic cube with a nonpalindromic cube root, among the cubes examined by Simmons, is 10,662,526,601. Its cube root, 2,201, had been noticed earlier by Trigg, who reported in 1961 that it was the only nonpalindrome with a palindromic cube less than 1,953,125,000,000. It is not yet known if 2,201 is the only integer with this property.

Simmons' computer search of palindromic fourth powers, to the same limit as his search of cubes, failed to uncover a single palindromic fourth power whose fourth root was not a palindrome of the general form 10 . . . 01. For powers 5 through 10 the computer found no palindromes at all except the trivial case of 1. Simmons conjectures that there are no palindromes of the form X^k where k is greater than 4.

"Repunits," numbers consisting entirely of 1's, produce palindromic squares when the number of units is one through nine, but 10 or more units give squares that are not palindromic. It has been erroneously stated that only primes have palindromic cubes, but this is disproved by an infinity of integers, the smallest of which is repunit 111. It is divisible by 3, yet its cube, 1,367,-631, is a palindrome. The number 836 is also of special interest. It is the largest three-digit integer whose square, 698,896, is palindromic, and 698,896 is the smallest palindromic square with an even number of digits. (Note also that the number remains palindromic when turned upside down.) Such palindromic squares are extremely rare. The next-larger one with an even number of digits is 637,832,238,736, the square of 798,644.

Turning to language palindromes, we first note that no common English words of more than seven letters are palindromic.

Examples of seven-letter palindromes are *reviver*, *repaper*, *deified* and *rotator*. The word "radar" (for radio detecting and ranging) is notable because it was coined to symbolize the reflection of radio waves. Dmitri Borgmann, whose files contain thousands of sentence palindromes in all major languages, asserts in his book *Language on Vacation* that the largest non-hyphenated word palindrome is *saippuakauppias*, a Finnish word for a soap dealer.

Among proper names in English, according to Borgmann, none is longer than Wassamassaw, a swamp north of Charleston, S.C. Legend has it, he writes, that it is an Indian word meaning "the worst place ever seen." Yreka Bakery has long been in business on West Miner Street in Yreka, Calif. Lon Nol, the former Cambodian premier, has a palindromic name, as does U Nu, once premier of Burma. Revilo P. Oliver, a classics professor at the University of Illinois, has the same first name as his father and grandfather. It was originally devised to make the name palindromic. If there is anyone with a longer palindromic name I do not know of it, although Borgmann suggests such possibilities as Norah Sara Sharon, Edna Lala Lalande, Duane Rollo Renaud, and many others.

There are thousands of excellent sentence palindromes in English, a few of which were discussed in a chapter on word play in my *Sixth Book of Mathematical Games from Scientific American*. The interested reader will find good collections in the Borgmann book cited above, and in the book by Howard Bergerson. Composing palindromes at night is one way for an insomniac to pass the dark hours, as Roger Angell so amusingly details in his article "Ainmosni" ("Insomnia" backward) in *The New Yorker*. I limit myself to one palindrome that is not well known, yet is remarkable for both its length and naturalness: "Doc note, I dissent. A fast never prevents a fatness. I diet on cod." It won a prize for James Michie in a palindrome contest sponsored by the *New Statesman* in England; results were published in the issue for May 5, 1967. Many of the winning palindromes are much longer than Michie's, but, as is usually the case, the longer palindromes are invariably difficult to understand.

Palindromists have employed various devices to make the unintelligibility of long palindromes more plausible: presenting them as telegrams, as one side only of a telephone conversation, and so on. Leigh Mercer, a leading British palindromist (he is the inventor of the famous "A man, a plan, a canal—Panama!"), has suggested a way of writing a palindrome as long as one wishes. The sentence has the form, " '———,' sides reversed, is '———.' " The first blank can be any sequence of letters, however long, which is repeated in reverse order in the second blank.

Good palindromes involving the names of U.S. presidents are exceptionally rare. Borgmann cites the crisp "Taft: fat!" as one of the shortest and best. Richard Nixon's name lends itself to "No 'x' in 'Mr. R. M. Nixon'?" although the sentence is a bit too contrived. A shorter, capitalized version of this palindrome, NO X IN NIXON, is also invertible.

The fact that "God" is "dog" backward has played a role in many sentence palindromes, as well as in orthodox psychoanalysis. In *Freud's Contribution to Psychiatry* A. A. Brill cites a rather farfetched analysis by Carl Jung and others of a patient suffering from a ticlike upward movement of his arms. The analysts decided that the tic had its origin in an unpleasant early visual experience involving dogs. Because of the "dog-god" reversal, and the man's religious convictions, his unconscious had developed the gesture to symbolize a warding off of the evil "dog-god." Edgar Allan Poe's frequent use of the reversal words "dim" and "mid" is pointed out by Humbert Humbert, the narrator of Vladimir Nabokov's novel *Lolita*. In the second canto of *Pale Fire*, in Nabokov's novel of the same title, the poet John Shade speaks of his dead daughter's propensity for word reversals:

> . . . *She twisted words: pot, top,*
> *Spider, redips. And "powder" was "red wop."*

Such word reversals, as well as sentences that are different sentences when they are spelled backward, are obviously close cousins of palindromes, but the topic is too large to go into here.

Palindrome sentences in which words, not letters, are the units have been a specialty of another British expert on word play, J. A. Lindon. Two splendid examples, from scores that he has composed, are:

"You can cage a swallow, can't you, but you can't swallow a cage, can you?"

"Girl, bathing on Bikini, eyeing boy, finds boy eyeing bikini on bathing girl."

Many attempts have been made to write letter-unit palindrome poems, some quite long, but without exception they are obscure, rhymeless, and lacking in other poetic values. Somewhat better poems can be achieved by making each line a separate palindrome rather than the entire poem, or by using the word as the unit. Lindon has written many poems of both types. A third type of palindrome poem, invented by Lindon, employs lines as units. The poem is unchanged when its lines are read forward but taken in reverse order. One is allowed, of course, to punctuate duplicate lines differently. The following example is one of Lindon's best:

> *As I was passing near the jail*
> *I met a man, but hurried by.*
> *His face was ghastly, grimly pale.*
> *He had a gun. I wondered why*
> *He had. A gun? I wondered . . . why,*
> *His face was ghastly! Grimly pale,*
> *I met a man, but hurried by,*
> *As I was passing near the jail.*

This longer one is also by Lindon. Both poems appear in Howard W. Bergerson's *Palindromes and Anagrams* (Dover, 1973).

DOPPELGÄNGER

Entering the lonely house with my wife,
*   I saw him for the first time*
Peering furtively from behind a bush—
*   Blackness that moved,*
*   A shape amid the shadows,*
A momentary glimpse of gleaming eyes
*   Revealed in the ragged moon.*
A closer look (he seemed to turn) might have
Put him to flight forever—
*   I dared not*
(For reasons that I failed to understand),
*   Though I knew I should act at once.*

I puzzled over it, hiding alone,
Watching the woman as she neared the gate.
*   He came, and I saw him crouching*
*   Night after night.*
*   Night after night*
*   He came, and I saw him crouching,*
Watching the woman as she neared the gate.

I puzzled over it, hiding alone—
*   Though I knew I should act at once,*
For reasons that I failed to understand
*   I dared not*
*   Put him to flight forever.*

A closer look (he seemed to turn) might have
*   Revealed in the ragged moon*
A momentary glimpse of gleaming eyes,
*   A shape amid the shadows,*
*   Blackness that moved.*

Peering furtively from behind a bush,
*   I saw him, for the first time,*
Entering the lonely house with my wife.

Lindon holds the record for the longest word ever worked into a letter-unit palindrome. To understand the palindrome you must know that Beryl has a husband who enjoys running around his yard without any clothes on. Ned has asked him if he does this to annoy his wife. He answers: "Named undenominationally rebel, I rile Beryl? La, no! I tan. I'm, O Ned, nude, man!"

ADDENDUM

A. Ross Eckler, editor and publisher of *Word Ways*, a quarterly journal on word play that has featured dozens of articles on palindromes of all types, wrote to say that the "palindromic gap" between English and other languages is perhaps not as wide as I suggested. The word "semitime" (in *Webster's Second*) can be pluralized to make a 9-letter palindrome, and "kinnikinnik" is in *Webster's Third*. Dmitri Borgmann pointed out in *Word Ways*, said Eckler, that an examination of foreign dictionaries failed to substantiate such long palindromic words as the Finnish soap dealer, suggesting that they are artificially created words.

Among palindromic towns and cities in the United States, Borgmann found the 7-letter Okonoko (in West Virginia). If a state (in full or abbreviated form) is part of the palindrome, Borgmann offers Apollo, Pa., and Adaven, Nevada. Some U.S. towns, Eckler continued, are intentional reversal pairs, such as Orestod and Dotsero, in Eagle County, Colorado, and Colver and Revloc, in Cambria County, Pennsylvania. Nova and Avon, he added, are Ohio towns that are an unintentional reversal pair.

George L. Hart III sent the following letter, which was published in *Scientific American*, November 1970:

Sirs:

Apropos of your discussion of palindromes, I would like to offer an example of what I believe to be the most complex and exquisite type of palindrome ever invented. It was devised by the Sanskrit aestheticians, who termed it *sarvatobhadra*, that is, "perfect in every direction." The most famous example of it is found in the epic poem entitled *Śiśupālavadha*.

```
sa - kā - ra - nā - nā - ra - kā - sa -
kā - ya - sā - da - da - sā - ya - kā
ra - sā - ha - vā   vā - ha - sā - ra -
nā - da -vā - da - da - vā - da - nā.
(nā da vā da da vā da nā
 ra sā ha vā vā ha sā ra
 kā ya sā da da sā ya kā
 sa kā ra nā nā ra kā sa)
```

Here hyphens indicate that the next syllable belongs to the same word. The last four lines, which are an inversion of the first four, are not part of the verse but are supplied so that its properties can be seen more easily. The verse is a description of an army and may be translated as follows: "[That army], which relished battle [rasāhavā], contained allies who brought low the bodes and gaits of their various striving enemies [sakāranānārakāsakāyasādadasāyakā], and in it the cries of the best of mounts contended with musical instruments [vāhasāranāda-vādadavādanā]."

Two readers, D. M. Gunn and Rosina Wilson, conveyed the sad news that the Yreka Bakery no longer existed. However, in 1970 its premises were occupied by the Yrella Gallery, and Ms. Wilson sent a Polaroid picture of the gallery's sign to prove it. Whether the gallery is still there or not, I do not know.

ANSWERS

READERS WERE ASKED to prove that no prime except 11 can be a palindrome if it has an even number of digits. The proof exploits a well-known test of divisibility by 11 (which will not be proved here): If the difference between the sum of all digits in even positions and the sum of all digits in odd positions is zero or a multiple of 11, the number is a multiple of 11. When a palindrome has an even number of digits, the digits in odd positions necessarily duplicate the digits in even positions; therefore the difference between the sums of the two sets must be zero. The palindrome, because it has 11 as a factor, cannot be prime.

The same divisibility test applies in all number systems when the factor to be tested is the system's base plus one. This proves that no palindrome with an even number of digits, in any number system, can be prime, with the possible exception of 11. The number 11 is prime if the system's base is one less than a prime, as it is in the decimal system.

CHAPTER 20

Dollar Bills

A REMARKABLE VARIETY of small man-made objects lend themselves to tricks and puzzles that are sometimes mathematical in character. Let's take a not-so-serious look at some puzzling aspects of dollar bills.

A curious folding stunt involving symmetry operations on the rectangular shape of a bill is well known to magicians. The performer holds the bill at each end, with the picture of Washington upright [see Figure 103]. He folds the bill in half lengthwise, then in half to the left, and once again in half to the left. Then he unfolds the bill, apparently by reversing the three previous steps, but now Washington is upside down! When others try to do the same thing, the bill stubbornly refuses to invert.

The secret lies in the second fold. Note that it is made by carrying the right half of the bill *behind* the left half. The third fold is made the opposite way. When those two folds are undone, they are both opened to the front. This has the effect of rotating the bill 180 degrees around a vertical axis, as you will see by comparing the bill at step 2 with the bill at step 6. Even so, the final inversion comes as a surprise. One must practice until the three folds can be made smoothly and quickly. The unfolding should be slow and deliberate while you assert (ma-

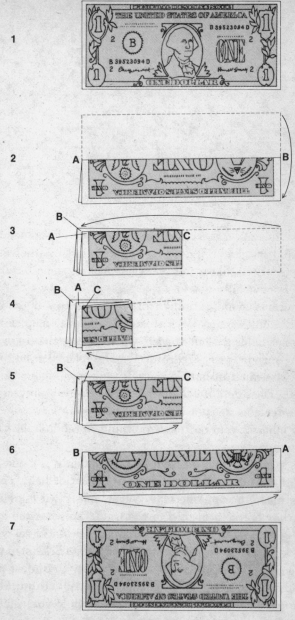

FIGURE 103
Inverting a dollar bill

gicians have the privilege of fibbing) that you are carefully repeating the same three steps in reverse time sequence.

Origami experts have devoted much time to devising ways of folding dollar bills into such things as a finger ring, bow tie, peacock, rabbit in hat; indeed, two treatises on this have been published by the Ireland Magic Company of Chicago: *The Folding Money Book* by Adolfo Cerceda (1963) and *The Folding Money Book Number Two* by Samuel and Jean Randlett (1968). The folds described in those books are rather complicated, but here is a simple one that readers are invited to discover for themselves. How can a dollar bill be given two creases in such a way that it makes the best possible picture of a mushroom?

All bills have identifying eight-digit serial numbers, and of course those numbers can figure in many different kinds of mathematical diversion. Has the reader ever played dollar-bill poker? Each of two players takes a bill from his pocket and the two then alternate claims of a pair or better, using the digits of the serial number as if they were cards. No straights or full houses are allowed, but sets of like digits may go higher than four of a kind. At each turn a player must raise his claim or call. Bluffing is permitted. After a call both numbers are inspected and the player who made the last claim is allowed to use the serial numbers on *both* bills to satisfy his claim. For example, if he had claimed six 3's and there are two 3's in his serial number and four or more in his opponent's, he wins his opponent's dollar. Otherwise he loses his dollar.

A trick that was a favorite of Royal V. Heath, a New York stockbroker and amateur magician who in 1933 wrote a book called *Mathemagic*, begins by someone's being asked to take a dollar bill from his pocket and look at the serial number. He calls out the sum of the first and second digits and then the sum of the second and third digits, the third and fourth, and so on to the end. For an eighth and final sum he adds the last and *second* digits. The performer jots down these eight sums as they are called. Without making any written calculations he immediately writes out the bill's serial number.

The problem is one of solving quickly a set of eight simultaneous equations. The solution goes back to Diophantus, a third-century algebraist who lived in Alexandria; the earliest presentation of it as a trick is in *Problèmes Plaisants et Délectables,* by Claude Gaspar Bachet (1612), Problem VII. There is a simple procedure for calculating the original number. Add the second, fourth, sixth, and eighth sums; subtract the sum of the third, fifth, and seventh sums; halve the result. This can be done easily in the head as the sums are called out. Starting with the second sum, the numbers are alternately subtracted and added as shown schematically in Figure 104. Halving the final result gives the second digit of the serial number. Instead of calling it out, however, the performer subtracts it from the first sum so that he can write and call out the serial number's first digit. It is a simple matter to give the remaining digits in order. The second is already known. Subtracting it from the second sum gives the third digit of the serial number. The third digit subtracted from the third sum gives the fourth digit, and so on to the end.

The trick is not limited to the digits of eight-digit numbers.

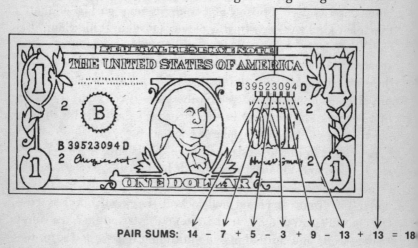

PAIR SUMS: 14 − 7 + 5 − 3 + 9 − 13 + 13 = 18

$\dfrac{18}{2}$ = 9 (SECOND DIGIT OF SERIAL NUMBER)

FIGURE 104

An ancient formula solves a dollar-bill trick

It applies to any series of real numbers, positive or negative, rational or irrational. If there is an even number of numbers in the series, the procedure just explained is used. If there is an odd number of numbers, you ask that the final sum be that of the last and *first* numbers. Instead of ignoring the first sum, start with the first sum and alternately subtract and add. Halving the final result then gives the first (not the second) number in the original series. Suppose, for instance, the series is 100, −27, 2/3, −1, 2,456. The five sums will be 73, −26⅓, −1/3, 2,455, and 2,556. When these are alternately subtracted and added, the result is 200. Half of 200 is 100, the first number of the original series. (For a puzzle based on all this, see Knot IV of Lewis Carroll's *A Tangled Tale*.)

Many tricks with serial numbers are based on what magicians call the "nine principle," which in turn derives from the fact that our number system is based on 10. For example, ask someone to take a bill from his pocket while you stand with your back turned. Have him jot down its serial number. Tell him to scramble the eight digits—that is, to write them down in any order—to make a second eight-digit number and then to subtract the smaller number from the larger. With your back still turned, ask him to cross out any digit (except zero) in the answer, then call out to you, in any order, the remaining digits. You immediately name the digit he crossed out.

The secret lies in the fact that if any number is scrambled and the smaller number taken from the larger, the difference has a digital root of 9. An example will make this clear. Suppose the serial number is 06281377 and this is scrambled to 87310267. The difference is 81028890. The digital root of this number is obtained by adding the digits in any order, casting out nines as you go along. Eight plus 1 plus 2 is 11, but in your mind you add the digits of 11 and remember only 2, which is the same as taking 9 from 11. Continue in this way, adding the digits whenever a partial sum has more than one digit. The single digit at the finish is the digital root. Since the number obtained by subtraction is certain to have a digital root of 9, it is easy to determine the missing digit. Merely add the digits as they are called, casting out nines as you go along. If the final digit is 9, your subject must have crossed out 9. Otherwise take the final digit from 9 to get the crossed-out digit.

Many other procedures also result in numbers with digital roots of 9. For instance, your subject can add the digits of the serial number, then subtract the sum from the serial number. Or he can add the digits, multiply by 8, and add the product to the original number. Instead of naming a number crossed out in the final result, you can calculate a person's age by asking him to add his age to the final result and call off, in any order, the digits in the sum. What you do is obtain the digital root of the numbers called out, then keep adding nines to it mentally until you reach what you estimate to be his age. Suppose a woman follows any of the above procedures that produce a number with a digital root of 9. She adds her age and gives you, in scrambled order, the digits in the sum. Assume they have a digital root of 4. In your mind you simply tick off: 4—13—22—31—40—49 and so on, picking the number that seems most likely to be her age.

For another trick based on the nine principle, obtain a bill with a serial number that has a digital root of 9 and carry it with you. You ask someone to jot down eight random digits but, before he starts, you appear to have an afterthought. Take out your bill and tell him to use the digits of its serial number; it is a handy way, you explain, to obtain random digits. While your back is turned he can scramble the number and *add* the two eight-digit numbers, then you can proceed with any of the tricks described above. Indeed, he can form as many scrambled eight-digit numbers as he wishes; their sum will always have a digital root of 9. He can scramble and multiply by any number; the product will have a digital root of 9. If he suspects that your bill is a special one and insists on using one of his own, shift to one of the previously described procedures.

A bit harder to puzzle out is the following serial-number stunt by Ben B. Braude that appeared in a magic periodical with the unlikely name *The Pallbearers Review* (see October 1967, page 127; December 1967, page 144). The subject writes down the serial number of his own bill, reverses it, and adds the two numbers. He crosses out any digit and reads the result aloud, substituting x for the missing digit. Suppose the serial number is 30956714. The subject begins by adding this to its reversal, 41765903. He crosses out 6 in the sum and reads aloud: $72722x17$. How can you figure the missing digit?

Serial numbers serve, of course, to forestall various kinds of criminal activity, but there is at least one swindle in which a bill's serial number plays an essential role. The fraud pops up from time to time in American bars. A man at one end of the bar starts doing magic tricks for the bartender and customers seated nearby. After a few tricks he announces that he will perform the most sensational trick he knows. It requires a $10 bill. He asks the bartender to lend him such a bill and, to make sure the identical bill is returned, asks him to copy down the bill's serial number. The magician folds this bill and apparently seals it in an envelope. Actually it is passed through a slot in the back of the envelope and palmed. The empty envelope is burned in an ashtray, seemingly destroying the bill. While the envelope burns, the magician secretly passes the bill to a confederate as he walks by on his way to the other end of the bar. The confederate uses the bill to pay another bartender for a drink. After the envelope is burned the magician tells the bartender to look in his cash register, where he will find the original bill. The bill is found, its serial number is checked, everyone is flabbergasted, and the two swindlers leave with a profit of about $9.

Does the reader know how to use a dollar bill as a ruler? The distance from the right side of the shield below the eagle to the right margin of the bill is one inch. The width of "United States" at the top of the green side is two inches. The rectangle containing the words "Federal Reserve Note" at the top of the bill's face is three inches wide. The bill itself is three-sixteenths of an inch longer than six inches. Eliminate one margin and you come very close to six inches.

I conclude with a series of puzzles, all concerning a $1 bill unless otherwise specified:

1. The numeral 1 appears in 10 places on a dollar bill, not counting those numbers that vary from bill to bill but including the 1 starting the year of the series and the Roman numeral I below the pyramid. How many times does a *word* for 1 appear?

2. How many times does the word "ten" appear on a $10 bill?

3. Find the date 1776 on a dollar bill.

4. Find a picture of a door key.

5. Find a word that is an anagram of "poetics."

6. Find a word that is an anagram of "a night snow."

7. Find these four-letter words: "sofa," "dose," "shin," "oral," "eats," "fame," "isle," "loft."

8. Find "Esau" and "Iva."

9. Find the phrase "at sea."

10. Find a Spanish word printed upside down.

11. Find a word with "O" as one of its letters, but an "O" pronounced like a "W."

12. What is the meaning of the eye above the pyramid and who suggested that it be put there?

13. On a $5 bill find "New Jersey" and the number 172.

14. If a $5 bill is tossed into the air, what is the probability that it will land with Lincoln's picture on the top side?

ANSWERS

THE FIRST PROBLEM was to fold a bill twice and produce a mushroom. It is done as in Figure 105.

The second problem concerned the sum of a bill's serial number and its reversal. When any number with an even number of digits is added to its reversal, the sum is always a multiple of 11. And all multiples of 11 have the following property: either the sum of the digits in the odd positions equals the sum of the

FIGURE 105
How to fold a $1 bill to make a mushroom

digits in the even positions or the sums differ by a multiple of 11. This provides a technique for determining the digit omitted from the sum of the serial number and its reversal. Simply obtain the sum of the even-position digits and the sum of the odd-position digits, then give x (the omitted digit) a value that will make the difference between those two sums either 0 or a multiple of 11. In the example the spectator calls out 72722x17. The odd-position digits here add to 17, the even-position digits to 11. Since x belongs to the even set, x must have a value that will raise 11 to 17. Therefore x equals 6. If the set containing x has a sum greater than 17, say 19, you add 11 to 17, making 28, then subtract 19 to arrive at 9 for the missing digit. (Alternatively, you could subtract 11 from 19 to get 8, then take 8 from 17 to arrive at 9.) If the set containing x has a sum that is less than the other sum and differs from it by more than 11, add 11 and subtract. If the sums of the two alternating sets are equal, the missing digit is 0.

Answers to the short questions follow:

1. A word for "1" appears nine times on a $1 bill. Did you overlook "unum"?

2. The word "ten" appears 13 times on a $10 bill. Did you overlook "ten" in "*ten*der" and "sep*tent*"?

3. The date 1776 appears in Roman numerals at the base of the pyramid.

4. The door key is in the green seal on the bill's face.

5. The anagram of "poetics" is "coeptis" (above the pyramid).

6. "Washington" is the anagram of "a night snow."

7. "Sofa," appears in "United States of America," "dose" in the Latin phrase below the pyramid, "shin" in "Washington," "oral" in "for all debts," "eats" in "great seal," "fame" in "of America," "isle" in "is legal," "loft" in "great seal of the."

8. "Esau" is in "Thesaur" (in the green seal). "Iva" is in "private."

9. "At sea" is in "great seal."

10. The inverted Spanish word is "si" (in "This note is . . .") and elsewhere. Reader Scott Brown found four others: "o," "no," "ni," and "os."

11. "One" contains "O" pronounced as a "W."

12. The eye above the pyramid is the "Eye of Providence." It was proposed by Benjamin Franklin to emphasize that the Union, symbolized by the 13-step pyramid, should always be under the watchful eye of God.

13. On a $5 bill "New Jersey" is the state name above the third and fourth columns of the Lincoln Memorial. You'll need a magnifying glass to see it. The number 172 can be seen as large dark numerals in the foliage at the base of the memorial, on the left. The number can be taken as 3172, but the 3 is not as distinct as the other numerals are.

14. The probability is 1. On the back of a $5 bill you will see Lincoln's statue inside the Lincoln Memorial.

Bibliography

1. OPTICAL ILLUSIONS

Optical Illusions. S. Tolansky. Macmillan, 1964.

Visual Illusions: Their Causes, Characteristics, and Applications. Matthew Luckiesh, with a new introduction by William H. Ittelson. Dover, 1965.

Eye and Brain. Richard L. Gregory. McGraw-Hill, 1966.

The Intelligent Eye. Richard L. Gregory. McGraw-Hill, 1970.

"Impossible Objects as Nonsense Sentences." D. A. Huffman in *Machine Intelligence 6.* Bernard Meltzer and Donald Michie (editors). Edinburgh University Press, 1971.

Illusion in Nature and Art. Richard L. Gregory and E. H. Gombrich. Scribner's, 1973.

"The Theory of Braids and the Analysis of Impossible Figures." Thaddeus M. Cowan, *Journal of Mathematical Psychology*, Vol. 11, August 1974, pages 190–212.

"Organizing the Properties of Impossible Figures." Thaddeus M. Cowan, *Perception*, Vol. 6, 1977, pages 41–56.

"Visual Illusions that Can Be Achieved by Putting a Dark Filter over One Eye." Jearl Walker. *Scientific American*, March 1978, Amateur Science Department.

2. MATCHES

Match-Stick Magic. Will Blyth. C. Arthur Pearson, 1921.

Match-ic. Martin Gardner. Ireland Magic Company, 1935.

"Match Puzzles." Henry Ernest Dudeney in *536 Puzzles and Curious Problems.* Martin Gardner (editor). Scribner's, 1967, pages 201–8.

"Hit-and-Run on a Graph." Jurg Nievergelt and Steve Chase, *Journal of Recreational Mathematics*, Vol. 1, April 1968, pages 112–17.

3. SPHERES AND HYPERSPHERES

An Introduction to the Geometry of N Dimensions. D. M. Y. Sommerville. Methuen & Co. Ltd., 1919; Dover, 1958.

"The Problem of the Thirteen Spheres." John Leech, *Mathemetical Gazette*, Vol. 40, February 1956, pages 22–23.

"On a Theorem in Geometry." Daniel Pedoe, *American Mathematical Monthly*, Vol. 74, June 1967, pages 627–40.

"Some Sphere Packings in Higher Space." John Leech, *Canadian Journal of Mathematics*, Vol. 16, 1964, pages 657–82.

"Notes on Sphere Packings." John Leech, *Canadian Journal of Mathematics*, Vol. 19, 1967, pages 251–67.

"Five Dimensional Non-Lattice Sphere Packings." John Leech, *Canadian Mathematical Bulletin*, Vol. 10, 1967, pages 387–93.

Twelve Geometrical Essays, Chapters 8, 9, and 12. H. S. M. Coxeter. Southern Illinois University Press, 1968.

"Six and Seven Dimensional Non-Lattice Sphere Packings." John Leech, *Canadian Mathematical Bulletin*, Vol. 12, 1969, pages 151–55.

"The Kiss Precise." W. S. Brown, *American Mathematical Monthly*, Vol. 76, June 1969, pages 661–63.

"New Sphere Packings in Dimensions 9–15." John Leech and N. J. A. Sloane, *Bulletin of the American Mathematical Society*, Vol. 76, September 1970, pages 1006–10.

"Sphere Packings and Error-Correcting Codes." John Leech and N. J. A. Sloane, *Canadian Journal of Mathematics*, Vol. 23, 1971, pages 718–45.

4. PATTERNS OF INDUCTION

A History of Board Games Other than Chess. H. J. R. Murray. Oxford University Press, 1952.

Games of the Orient. Stewart Culin. Charles Tuttle, 1958.

Board and Table Games from Many Civilizations. R. C. Bell. Oxford University Press, 1960.

Board and Table Games 2 from Many Civilizations. R. C. Bell. Oxford University Press, 1969.

A Gamut of Games. Sidney Sackson. Random House, 1969.

Games of the World. Frederic V. Grunfeld, editor. Holt, Rinehart and Winston, 1975.

5. ELEGANT TRIANGLES

College Geometry: An Introduction to the Modern Geometry of the Triangle and the Circle. Nathan Altshiller Court. Barnes and Noble, 1952.

Geometry Revisited. H. S. M. Coxeter and Samuel L. Greitzer. Random House, 1967.

Fourth-equilateral-triangle problem

Introduction to Geometry. H. S. M. Coxeter. Wiley, 1961, Section 1.8.

Geometric Transformations. I. M. Yaglom. Random House, 1962, pages 93–94.

"The Triangle Reinvestigated." J. Garfunkel, *American Mathematical Monthly*, Vol. 72, January 1965, pages 12–20.

Trisecting-cevians problem

One Hundred Mathematical Curiosities. William R. Ransom. Walch, 1955, pages 172–73.

Ingenious Mathematical Problems and Methods. L. A. Graham. Dover, 1959, Problem 52.

Introduction to Geometry. H. S. M. Coxeter. Wiley, 1961, Section 13.55.

Mathematical Snapshots. Hugo Steinhaus. Oxford University Press, third revised edition, 1969, pages 8–9.

"Trisection Triangle Problems." Marjorie Bicknell, *Mathematics Magazine*, Vol. 69, February 1976, pages 129–34.

Five-con triangles

"More-than-similar Triangles." C. Salkind, *Mathematics Teacher*, Vol. 47, December 1954, pages 561–62.

"Problem E 1162." Victor Thebault, *American Mathematical Monthly*, Vol. 62, 1955, pages 255, 729–30.

Ingenious Mathematical Problems and Methods. L. A. Graham. Dover, 1959, Problem 59.

"Mystery Puzzler and Phi." Marvin H. Holt, *Fibonacci Quarterly*, Vol. 3, April 1965, pages 135–38.

"5-Con Triangles." Richard G. Pawley, *Mathematics Teacher*, Vol. 60, May 1967, pages 438–43.

Fibonacci and Lucas Numbers. Verner E. Hoggatt, Jr. Houghton Mifflin, 1969, Chapter 4.

"Almost Congruent Triangles." Robert T. Jones and Bruce B. Peterson, *Mathematics Magazine*, Vol. 47, September 1974, pages 180–89.

Heron's formula

A Short Account of the History of Mathematics. W. W. Rouse Ball. Dover, 1960, Chapter 4.

"A Simpler Proof of Heron's Formula." Claude H. Raifaizen, *Mathematics Magazine*, Vol. 44, January 1941, pages 27–28.

Three-distances problem

Ingenious Mathematical Problems and Methods. L. A. Graham. Dover, 1959, Problem 55.

Mathematical Quickies. Charles W. Trigg. McGraw-Hill, 1967, Problem 201.

The crossed-ladders problem

"Answer to Problem E 433." Albert A. Bennett, *American Mathematical Monthly*, Vol. 48, April 1941, pages 268–69.

Ingenious Mathematical Problems and Methods. L. A. Graham. Dover, 1952, Problem 25.

One Hundred Mathematical Curiosities. William R. Ransom. Walch, 1955, pages 43–46.

Mathematical Puzzles. Geoffrey Mott-Smith. Dover, 1954, Problem 103.

"The Crossed Ladders." H. A. Arnold, *Mathematics Magazine*, Vol. 29, 1956, pages 153–54.

"Complete Solution of the Ladder Problem in Integers." Alan Sutcliffe, *Mathematical Gazette*, Vol. 47, May 1963, pages 133–36.

"Answer to Problem 5323." Gerald J. Janusz, *American Mathematical Monthly*, Vol. 73, December 1966, pages 1125–27.

The Surprise Attack in Mathematics. L. A. Graham. Dover, 1968, Problem 6.

"The Ladder Problem." William L. Schaaf in *A Bibliography of Recreational Mathematics, Vol. 1*. National Council of Teachers of Mathematics, fourth edition, 1970, pages 31–32.

6 AND 7. RANDOM WALKS

"Random Paths in Two and Three Dimensions." W. H. McCrea and F. J. Whipple, *Proceedings of the Royal Society of Edinburgh*, Vol. 60, 1940, pages 281–98.

An Introduction to Probability Theory and Its Applications. William Feller. Wiley, Vol. 1, 1957; Vol. 2, 1966.

"Finite Markov Chains." John G. Kemeny, Hazleton Mirkil, J. Laurie Snell, and Gerald L. Thompson in *Finite Mathematical Structures*. Prentice-Hall, 1959.

Probability and Statistics. Frederick Mosteller, Robert E. K. Rourke, and George B. Thomas, Jr. Addison-Wesley, 1961.

Random Walks. E. B. Dynkin and V. A. Uspenskii. Heath, 1963.

"Random Walks." John G. Kemeny in *Enrichment in Mathematics for High School*, Chapter 21. National Council of Teachers of Mathematics, 1963.

Principles of Random Walk. Frank Spitzer. Van Nostrand, 1964.

The Theory of Gambling and Statistical Logic. Richard A. Epstein. Academic Press, 1967.

"Random Hops on Polyhedral Edges and Other Networks." R. Robinson Rowe. *Journal of Recreational Mathematics*, Vol. 4, April 1971, pages 124–34; see also pages 142–43.

8. BOOLEAN ALGEBRA

Applied Boolean Algebra. Franz E. Hohn. Macmillan, 1960.

Boolean Algebra and Its Applications. J. Eldon Whitesitt. Addison-Wesley, 1961.

Logic and Boolean Algebra. B. H. Arnold. Prentice-Hall, 1962.

Boolean Algebra. R. L. Goodstein. Pergamon Press, 1963.

"George Boole (1815–1864)." T. A. A. Broadstreet, *Mathematical Gazette,* Vol. 48, December 1964, pages 373–78.

"In Praise of Boole." Norman T. Gridgeman, *New Scientist,* December 3, 1964, pages 655–57.

Sets, Lattices, and Boolean Algebras. James C. Abbott. Allyn and Bacon, 1969.

Boolean Algebras. Roman Sikorski. Springer-Verlag, third edition, 1969.

Boolean Algebra and Its Uses. G. F. South. Van Nostrand, 1974.

9. CAN MACHINES THINK?

"Computing Machinery and Intelligence." A. M. Turing, *Mind,* Vol. 59, October 1950, pages 433–60.

Dimensions of Mind. Sidney Hook (editor). New York University Press, 1960.

"The Supercomputer as Liar." Michael Scriven, *British Journal for the Philosophy of Science,* Vol. 13, February 1963, pages 313–15.

"The Imitation Game." Keith Gunderson, *Mind,* Vol. 73, April 1964, pages 234–45.

Minds and Machines. Alan Ross Anderson (editor). Prentice-Hall, 1964.

The Difference of Man and the Difference it Makes. Mortimer J. Adler. Holt, Rinehart and Winston, 1967. (See review by Anthony Quinton in the *New York Review of Books,* November 21, 1968.)

Mentality and Machines. Keith Gunderson. Doubleday Anchor, 1971.

"Turing Machines and the Mind-Body Problem." J. J. Clarke, *British Journal for the Philosophy of Science,* Vol. 23, February 1972, pages 1–12.

What Computers Can't Do. Hubert L. Dreyfus. Harper and Row, 1972. (For an opposite point of view see *The Artificial Intelligence of Hubert L. Dreyfus,* by Seymour Papert, M.I.T. Project Mac, Artificial Intelligence Memo No. 154, January 1968; and my review of Dreyfus' book in *Book World,* January 23, 1968.)

Computer Power and Human Reason. Joseph Weizenbaum. W. H. Freeman, 1976.

10. CYCLIC NUMBERS

"On Cyclic Numbers." Solomon Guttmann, *American Mathematical Monthly,* Vol. 41, March 1934, pages 159–66.

Cycles of Recurring Decimals, Vols. 1 and 2. D. R. Kaprekar. India: privately published, 1950, 1953.

"The Cyclic Number 142857." C. A. Costa Aliago, *Scripta Mathematica,* Vol. 19, 1953, pages 181–84.

"Primes and Recurring Decimals." R. E. Green, *Mathematical Gazette,* Vol. 47, February 1963, pages 25–33.

"Cyclic Toward Infinity." Albert H. Beiler. *Recreations in the Theory of Numbers.* Dover, 1964, Chapter 10.

"Full-Period Primes." Samuel Yates, *Journal of Recreational Mathematics*, Vol. 3, October 1970, pages 221–24.

"9:8:7." Samuel Yates, *Journal of Recreational Mathematics*, Vol. 8, 1975, pages 279–80.

"Numeroddities." Clifford C. Corbett, *Journal of Recreational Mathematics*, Vol. 8, 1975, pages 218–22.

Prime Period Lengths. Samuel Yates. Privately published in 1975 by the author, 104 Brentwood Drive, Mt. Laurel, N.J. 08057. Gives the periodic lengths of reciprocals of all primes from 3 through 1,370,471.

"Some Amazements of Mathematics." Charles S. Peirce in *The New Elements of Mathematics*. Carolyn Eisele (editor). Humanities Press, 1976, Vol. 3, Part I, pages 593–604.

Magic and lightning-calculation tricks based on cyclic numbers

Les Calculateurs Prodiges. D. Jules Regnault. Paris: Payot, 1943, pages 336–43. Lightning-calculation tricks based on the periods for 1/7 and 1/19.

Math Miracles. Wallace Lee. Durham, N.C.: privately published, pages 77–78. Lightning-calculation trick based on period for 1/17.

Mathematics, Magic and Mystery. Martin Gardner. Dover, 1956. Card trick based on 142,857.

The Magic of Numbers. Robert Toquet. A. S. Barnes, 1961; Premier paperback, 1962, pages 54–57. Lightning-calculation trick based on the period of 1/19.

Mathematical Magic. William Simon. Scribner's, 1964, pages 31–35. A slate prediction trick based on 142,857.

"Numbers." L. Vosburgh Lyons, *The Pallbearers Review*, December 1969, pages 310 ff. An ingenious stage trick using numbered blocks, a pair of dice, secret magnets, and the period of 1/13.

12. DOMINOES

Récréations Mathématiques. Edouard Lucas. Paperback reprints, Paris, 1960, Vol. 2, pages 52–63; Vol. 4, pages 123–51.

"Quadrilles." Wade E. Philpott, *Recreational Mathematics Magazine*, January–February 1964, pages 5–11.

"Domino Recreations." Joseph Madachy, in *Mathematics on Vacation*, pages 209–219. Scribner's, 1966.

"A General Quadrille Solution." Wade E. Philpott, *Mathematical Gazette*, December 1967, pages 287–90.

"Domino Puzzles." Henry Ernest Dudeney in *536 Puzzles and Curious Problems*. Martin Gardner (editor). Scribner's, 1967, pages 189–98.

"Some Domino Puzzles." Frederik Schuh in *The Master Book of Mathematical Recreations*. T. H. O'Beirne (editor). Dover, 1968, Chapter 2.

"Domino Games." R. C. Bell, *Board and Table Games*, Oxford University Press, 1960, Chapter 6; *Board and Table Games 2*, Oxford University Press, 1969, Chapter 6.

"Domino and Superdomino Recreations," Parts 1–5. Wade E. Philpott in *Journal of Recreational Mathematics*, January, April, October 1971; April, July 1972.

Domino Games and Domino Puzzles. Karel W. H. Leeflang. St. Martin's Press, 1972.

The Domino Book. Frederick Berndt. Thomas Nelson, 1974; Bantam paperback, 1975.

13. FIBONACCI AND LUCAS NUMBERS

Fibonacci Numbers. N. N. Vorob'ev. Blaisdell, 1961.

"A Generalized Fibonacci Sequence." A. F. Horadam, *American Mathematical Monthly*, Vol. 68, May 1961, pages 455–59.

An Introduction to Fibonacci Discovery. Brother U. Alfred. The Fibonacci Association, 1965.

"The Fibonacci Sequence." J. H. Cadwell in *Topics in Recreational Mathematics*. Cambridge University Press, 1966, Chapter 2.

Fibonacci and Lucas Numbers. Verner E. Hoggatt, Jr. Houghton Mifflin, 1969.

"Fibonacci Numbers." Verner E. Hoggatt, Jr., *Britannica Yearbook of Science and the Future, 1977*. Encyclopaedia Britannica, 1976, pages 177–92.

"Fibonacci Fantasy." Martin Gardner, *Apocalypse* (a U.S. magic periodical), Vol. 1, August 1978, pages 88–89. Some mathematical magic tricks with numbers and playing cards based on a discovery by Stewart James about the digital roots of generalized Fibonacci sequences.

14. SIMPLICITY

"Two Measures of Simplicity." John G. Kemeny, *Journal of Philosophy*, Vol. 52, 1955, pages 722–33.

"The Test of Simplicity." Nelson Goodman, *Science*, Vol. 128, October 31, 1958, pages 1064–69.

The Logic of Scientific Discovery. Karl Popper. Basic Books, 1959, Chapter 7.

The Myth of Simplicity. Mario Bunge. Prentice-Hall, 1963.

"Simplicity." Carl G. Hempel in *Philosophy of Natural Science*. Prentice-Hall, 1966, pages 40–45.

"Simplicity." Mary Hesse in *The Encyclopedia of Philosophy*. Macmillan Company and the Free Press, 1967.

"Science and Simplicity." Nelson Goodman in *Philosophy of Science Today*, edited by Sidney Morgenbesser. Basic Books, 1967.

"Ramifications of 'Grue.'" Mary Hesse, *The British Journal for the Philosophy of Science*, Vol. 20, May 1969, pages 13–25.

Simplicity. Elliott Sober. Clarendon Press, 1975.

On Leo Moser's spot problem

Challenging Mathematical Problems with Elementary Solutions, Vol. 1.
A. M. Yaglom and I. M. Yaglom. Holden-Day, Inc., 1964, pages 108-12.

"The Quest for the Lost Region." John Glenn, *Mathematics Teaching*, Vol.
43, 1968, pages 23–25.

"Induction: Fallible but Valuable." Jay Graening, *Mathematics Teacher*,
February 1971, pages 127–31.

"The Dissection of a Circle by Chords." T. Murphy, *Mathematical Gazette*,
No. 396, 1972, pages 113–15.

"Euler, Pascal, and the Missing Region." Richard A. Gibbs. *Mathematics
Teacher*, January 1973, pages 27–28.

16. SOLAR SYSTEM ODDITIES

"The Wheel Paradox Controversy." *Scientific American*, Vol. 16, 1866, page
347, through Vol. 18, 1868, page 366.

"Cole's Rule for Planetary Distances." H. W. Gould, *Proceedings of the
West Virginia Academy of Science*, Vol. 37, 1965, pages 243–57.

"A Subjective Evaluation of Bode's Law and an 'Objective' Test for Approxi-
mate Numerical Rationality." I. J. Good, *Journal of the American Statistical Association*, Vol. 64, March 1969, pages 23-49.

"Fibonacci-Series in the Solar System." B. A. Read, *Fibonacci Quarterly*,
Vol. 8, 1970, pages 428–438.

"The Titius-Bode Law: A Strange Bicentenary." Stanley L. Jaki, *Sky and
Telescope*, Vol. 43, 1972, pages 280–81.

The Titius-Bode Law of Planetary Distances: Its History and Theory.
Michael Martin Nieto. Pergamon Press, 1973. Reviewed by Philip Morri-
son in *Scientific American*, September 1973, page 194.

17. MASCHERONI CONSTRUCTIONS

Geometrical Constructions with a Ruler Given a Fixed Circle with its Center.
Jacob Steiner. *Scripta Mathematica*, 1950.

"Mascheroni Constructions." Julius H. Hlavaty. *Mathematics Teacher*, November 1957, pages 482–87.

"Mascheroni Constructions." Nathan Altshiller Court. *Mathematics Teacher*,
May 1958, pages 370–72.

"The Geometry of the Fixed-Compass." Arthur E. Hallerberg, *Mathematics
Teacher*, April 1959, pages 230–44.

"Georg Mohr and Euclidis Curiosi." Arthur E. Hallerberg. *Mathematics
Teacher*, February 1960, pages 127–32.

Geometrical Constructions Using Compasses Only. A. N. Kostovskii. Blaisdell, 1961.

The Ruler in Geometrical Constructions. A. S. Smogorzhevskii. Blaisdell,
1961.

"Joseph Moxon, Mathematical Practitioner." Arthur E. Hallerberg, *Mathematics Teacher*, October 1962, pages 490–92.

18. THE ABACUS

History

Number Words and Number Symbols: A Cultural History of Numbers. Karl Menninger. M.I.T. Press, 1969.

The History of the Abacus. J. N. Pullan. Frederick A. Praeger, 1969.

The Abacus: Its History, Its Design, Its Possibilities in the Modern World. Parry Moon. Gordon & Breach, 1971.

The Japanese abacus

The Japanese Abacus. Takashi Kojima. Charles E. Tuttle, 1954.

Advanced Abacus. Takashi Kojima. Charles E. Tuttle, 1963.

The Japanese Abacus Explained. Y. Yoshino. Introduction by Martin Gardner. Dover, 1963.

"Addition and Subtraction on the Soroban." Victor E. Haas, *Mathematics Teacher*, November 1965, pages 608–20.

The Chinese abacus

The Chinese Abacus. F. C. Scesney, 1944.

The Principles and Practice of the Chinese Abacus. Chung-Chien Liu. Hong Kong, 1958.

19. PALINDROMES

The palindrome number conjecture

Recreation in Mathematics. Roland Sprague. Dover, 1963, Problem 5.

"Palindromes by Addition." Charles W. Trigg, *Mathematics Magazine*, Vol. 40, January 1967, pages 26–28.

"Palindromes by Addition in Base Two." Alfred Brousseau. *Mathematics Magazine*, Vol. 42, November 1969, pages 254–56.

"More on Palindromes by Reversal-Addition." Charles W. Trigg, *Mathematics Magazine*, Vol. 45, September 1972, pages 184–86.

"On Palindromes." Heiko Harborth, *Mathematics Magazine*, Vol. 46, March 1973, pages 96–99.

"Versum Sequences in the Binary System." Charles W. Trigg, *Pacific Journal of Mathematics*, Vol. 47, 1973, pages 263–75.

"Palindromes: For Those Who Like to Start at the Beginning." Walter Koetke, *Creative Computing*, January 1975, pages 10–12.

"Follow-up on Palindromes." David Ahl, *Creative Computing*, May 1975, page 18.

"The 196 Problem." *Popular Computing*, Vol. 3, September 1975, pages 6–9.

Palindromic primes

"Patterns in Primes." Leslie Card, *Journal of Recreational Mathematics*, Vol. 1, April 1968, pages 93–99.

"On Palindromes and Palindromic Primes." Hyman Gabai and Daniel Coogan, *Mathematics Magazine*, Vol. 42, November 1969, pages 252–54.

"Special Palindromic Primes." Charles W. Trigg, *Journal of Recreational Mathematics*, Vol. 4, July 1971, pages 169–70.

Palindromic powers

"Palindromic Cubes." Charles W. Trigg, *Mathematics Magazine*, Vol. 34, March 1961, page 214.

"Palindromic Powers." Gustavus J. Simmons, *Journal of Recreational Mathematics*, Vol. 3, April 1970, pages 93–98.

"On Palindromic Squares of Non-Palindromic Numbers." Gustavus J. Simmons, *Journal of Recreational Mathematics*, Vol. 5, January 1972, pages 11–19.

Word palindromes

Oddities and Curiosities of Words and Literature. C. C. Bombaugh. Martin Gardner (editor). Dover, 1961.

Language on Vacation. Dmitri Borgmann. Scribner's, 1965.

Beyond Language. Dmitri Borgmann. Scribner's, 1967.

Word Ways: The Journal of Recreational Linguistics. Vol. 1, February 1968 to date.

"Weekend Competition." *New Statesman*, April 14, 1967, page 521; May 5, 1967, page 630; December 22, 1972, page 955; March 9, 1973, page 355.

"Ainmosni." Roger Angell, *The New Yorker*, May 31, 1969, pages 32–34.

"The Contest." *Maclean's Magazine*, December 1969, pages 68–69.

"Results of Competition." *New York Magazine*, December 22, 1969, page 82.

"I Call on Professor Osseforp." Solomon W. Golomb, *Harvard Bulletin*, March 1972, page. 45.

Palindromes and Anagrams. Howard W. Bergerson. Dover, 1973.

"The Oulipo." Martin Gardner, *Scientific American*, February 1977, pages 121–26.

20. DOLLAR BILLS

Mathematics, Magic and Mystery. Martin Gardner. Dover, 1956.

The Folding Money Book. Adolfo Cerceda. Ireland Magic Company, 1963.

Mathematical Magic. William Simon. Scribner's, 1964.

The Folding Money Book Number Two. Samuel and Jean Randlett. Ireland Magic Company, 1968.